Florida

State Assessments
Grade 7
Mathematics
SUCCESS STRATEGIES

**FSA Test Review for the
Florida Standards Assessments**

Dear Future Exam Success Story:

Congratulations on your purchase of our study guide. Our goal in writing our study guide was to cover the content on the test, as well as provide insight into typical test taking mistakes and how to overcome them.

Standardized tests are a key component of being successful, which only increases the importance of doing well in the high-pressure high-stakes environment of test day. How well you do on this test will have a significant impact on your future, and we have the research and practical advice to help you execute on test day.

The product you're reading now is designed to exploit weaknesses in the test itself, and help you avoid the most common errors test takers frequently make.

How to use this study guide

We don't want to waste your time. Our study guide is fast-paced and fluff-free. We suggest going through it a number of times, as repetition is an important part of learning new information and concepts.

First, read through the study guide completely to get a feel for the content and organization. Read the general success strategies first, and then proceed to the content sections. Each tip has been carefully selected for its effectiveness.

Second, read through the study guide again, and take notes in the margins and highlight those sections where you may have a particular weakness.

Finally, bring the manual with you on test day and study it before the exam begins.

Your success is our success

We would be delighted to hear about your success. Send us an email and tell us your story. Thanks for your business and we wish you continued success.

Sincerely,

Mometrix Test Preparation Team

Need more help? Check out our flashcards at:
http://MometrixFlashcards.com/FSA

TABLE OF CONTENTS

Top 15 Test Taking Tips...1

Expressions and Equations 2

Statistics and Probability 9

Geometry...23

Ratios and Proportional Relationships...39

The Number System 50

Practice Test #1 59

    Practice Questions..59

    Answers and Explanations 73

Practice Test #2 78

    Practice Questions..78

    Answers and Explanations 92

Success Strategies 97

How to Overcome Test Anxiety..103

    Lack of Preparation 103

    Physical Signals 104

    Nervousness 104

    Study Steps...106

    Helpful Techniques 107

Additional Bonus Material..113

Top 15 Test Taking Tips

1. Know the test directions, duration, topics, question types, how many questions
2. Setup a flexible study schedule at least 3-4 weeks before test day
3. Study during the time of day you are most alert, relaxed, and stress free
4. Maximize your learning style; visual learner use visual study aids, auditory learner use auditory study aids
5. Focus on your weakest knowledge base
6. Find a study partner to review with and help clarify questions
7. Practice, practice, practice
8. Get a good night's sleep; don't try to cram the night before the test
9. Eat a well balanced meal
10. Wear comfortable, loose fitting, layered clothing; prepare for it to be either cold or hot during the test
11. Eliminate the obviously wrong answer choices, then guess the first remaining choice
12. Pace yourself; don't rush, but keep working and move on if you get stuck
13. Maintain a positive attitude even if the test is going poorly
14. Keep your first answer unless you are positive it is wrong
15. Check your work, don't make a careless mistake

Expressions and Equations

Commutative property of addition and multiplication

The commutative property of addition states that the order in which two numbers are added does not change their sum; the commutative property of multiplication states that the order in which two numbers are multiplied does not change their product.

$$a + b = b + a$$
$$ab = ba$$

Associative property of addition and multiplication

The associate property of addition states that a series of added numbers can be grouped in various ways without affecting the sum; the associative property of multiplication states that a series of multiplied numbers can be grouped in various ways without affecting the product.

$$a + (b + c) = (a + b) + c$$
$$a(bc) = (ab)c$$

Additive identity and the multiplicative identity

The additive identity is the number which can be added to a number without changing its value; that number is zero. The multiplicative identity is the number which can be multiplied by a number without changing its value; that number is one.

<u>Example</u>

Use the distributive property to simplify $-\frac{1}{2}(x - 8)$.

The distributive property states that $a(b + c) = ab + ac$ and $a(b - c) = ab - ac$.

$$-\frac{1}{2}(x - 8) = \left(-\frac{1}{2}\right)(x) - \left(-\frac{1}{2}\right)(8) = -\frac{1}{2}x + 4.$$

<u>Example</u>

Name the property used in each step of simplifying $\frac{3}{4} \cdot \frac{2}{3} + \frac{3}{4} \cdot \frac{1}{3}$.

$$\frac{3}{4} \cdot \frac{2}{3} + \frac{3}{4} \cdot \frac{1}{3} = \frac{3}{4}\left(\frac{2}{3} + \frac{1}{3}\right)$$

$$\frac{3}{4}\left(\frac{2}{3} + \frac{1}{3}\right) = \frac{3}{4}(1)$$

$$\frac{3}{4}(1) = \frac{3}{4}$$

$\frac{3}{4} \cdot \frac{2}{3} + \frac{3}{4} \cdot \frac{1}{3} = \frac{3}{4}\left(\frac{2}{3} + \frac{1}{3}\right)$	Distributive property
$\frac{3}{4}\left(\frac{2}{3} + \frac{1}{3}\right) = \frac{3}{4}(1)$	Substitution property of equality
	(Since $\frac{2}{3} + \frac{1}{3} = 1$, the number 1 can replace the expression $\frac{2}{3} + \frac{1}{3}$.)
$\frac{3}{4}(1) = \frac{3}{4}$	Multiplicative identity

<u>Example</u>

Simplify.

$$-\frac{2}{3}\left(a - \frac{1}{4}\right)$$

$$\left(\frac{3}{4}x + 4\right) + \left(\frac{1}{4}x - 3\right)$$

Use the distributive property to simplify $-\frac{2}{3}\left(a - \frac{1}{4}\right)$.

$$-\frac{2}{3}\left(a - \frac{1}{4}\right) = -\frac{2}{3} \cdot a + \left(-\frac{2}{3}\right)\left(-\frac{1}{4}\right)$$

$$= -\frac{2}{3}a + \frac{2}{12}$$

$$= -\frac{2}{3}a + \frac{1}{6}$$

Use the associative and commutative properties of addition to simplify $\left(\frac{3}{4}x + 4\right) + \left(\frac{1}{4}x - 3\right)$.

$$\left(\frac{3}{4}x + 4\right) + \left(\frac{1}{4}x - 3\right) = \frac{3}{4}x + 4 + \frac{1}{4}x - 3$$

$$= \frac{3}{4}x + \frac{1}{4}x + 4 - 3$$

$$= x + 1$$

Example

Factor $\frac{1}{2}$ from the expression $\frac{1}{4}x - \frac{1}{2}$.

To factor $\frac{1}{2}$ from the expression $\frac{1}{4}x - \frac{1}{2}$, divide each term in the expression by $\frac{1}{2}$.

$$\frac{1}{4}x \div \frac{1}{2} = \frac{1}{4} \cdot x \cdot \frac{2}{1} = \frac{1}{4} \cdot \frac{2}{1} \cdot x = \frac{2}{4} \cdot x = \frac{1}{2}x$$

$$-\frac{1}{2} \div \frac{1}{2} = -1$$

The factored expression is $\frac{1}{2}\left(\frac{1}{2}x - 1\right)$.

Words and/or phrases

Addition
Some words and phrases that indicate addition are sum, plus, total, and, increased by, more, together, added to, combined with, gain.

Subtraction
Some words and phrases that indicate subtraction are difference, minus, less, decreased by, take away, fewer than, from, subtracted from, loss.

Multiplication
Some words and phrases that indicate multiplication are product, times, multiplied by, of, twice/double (×2), thrice/triple (×3).

Division
Some words and phrases that indication division are quotient, divided by, into, among, between, over, per, for every, ratio of, out of.
These lists are not exhaustive.

Example

Joshua calculates that the product of $19\frac{3}{4}$ and $10\frac{1}{4}$ is $404\frac{7}{8}$. Use mental estimation to determine the reasonableness of his answer.

The product of two numbers is found by multiplying those numbers, so the product of $19\frac{3}{4}$ and $10\frac{1}{4}$ is about 200 since $19\frac{3}{4}$ is close to 20 and $10\frac{1}{4}$ is close to 10. An answer of $404\frac{7}{8}$ seems unreasonable, so Joshua should check his calculation.

<u>Example</u>

Suppose you wish to center a 3 ¼ ft wide painting over a buffet which is 5 ¾ ft wide. Approximate how far each edge of the painting would be from each edge of the buffet.

The painting is just over 3 ft wide, and the buffet is just under 6 feet wide. So, the difference in the width of the buffet and the painting is about 3 ft.. The painting is centered above the buffet, half of the difference in width will be space to the left of the painting and the other half will be space to the right of the painting. Half of 3 ft is 1 ½ ft. So, each edge of the painting should be about 1 ½ ft from each edge of the buffet.

Mathematical symbols

= equals, is equal to, is, was, were, will be, yields, is the same as, amounts to, becomes
>**is** greater than, **is** more than
≥ **is** greater than or equal to, is at least, is no less than
<**is** less than, **is** fewer than
≤ **is** less than or equal to, is at most, is no more than

<u>Example</u>

Write three sentences which would translate into the inequality $2(x + 4) \geq 6$.

There are many ways to write $2(x + 4) \geq 6$ as a sentence, including
Two times the sum of a number and four is greater than or equal to six.
Twice the quantity x increased by four is at least six.
The product of two and a number to which four has been added must be no less than six.

Example

An $80 dress is marked down 25%. Find the price of the dress after the discount.

One way to find the price of the dress after the discount is to calculate the amount of the discount and subtract it from the price of the dress: The amount of the discount is 25% of $80, or $0.25 \times \$80 = \$20$. The price of the dress after the discount is $\$80 - \$20 = \$60$. Another way to find the discounted price is to write, simplify, and evaluate an expression representing the problem:

- If p = original price, then $0.25p$ represents the amount of the discount. So, an expression for the new price of the dress is $p - 0.25p$, which simplifies to $0.75p$. In other words, the discounted price of the dress is 75% of the original price. $0.75p = 0.75(\$80) = \$60$.

Example

If px represents the price of an item, write an expression can be used to find

- The price of the item with 8% sales tax.
- The pre-tax price of the item during a half-off sale.
- The after-tax price of the item during a half-off sale.

The amount of an 8% sales tax added to an item which costs px dollars is $0.08p$. The price of the item with the sales tax is $p + 0.08p$, which simplifies to $\mathbf{1.08p}$.

If the item is half off, the amount of the discount is $\frac{1}{2}p$. The new price of the item is the original price minus the discount, or $-\frac{1}{2}p$, which simplifies to $\frac{1}{2}\boldsymbol{p}$.

The price of the half-off item with 8% sales tax is $1.08\left(\frac{1}{2}p\right) = 1.08(0.5p) = \mathbf{0.54p}$.

Determining the order of operations used to simplify expressions

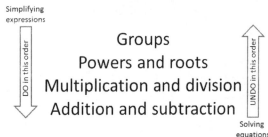

When simplifying an expression, work first within groups, which can be found within grouping symbols such as parentheses but also can be found under radical signs, in numerators or denominators of fractions, as exponents, etc. without such grouping symbols. Next, simplify powers and roots. Then multiply and divide from left to right; finally, add and subtract from left to right.

Relate this to the sequence of operations used when solving equations
When solving an equation, it is often helpful to first use order of operations to simplify the expressions on both sides of the equation, if possible. Then, undo the operations which have been performed on the variable by using inverse operations in reverse order of operations.

Example
Alina spent $20 at the fair. She paid $2 for admission plus $1.50 for every ride. Write and solve an equation to determine how many rides she rode.

Let x = the number of rides Alina rode at the fair. The expression $1.50x$ represents the amount of money spent riding rides. Alina spent a total of $1.50x + $2 on rides and admission; this amount equals $20, so

$$\begin{aligned} \$1.50x + \$2 &= \$20 \\ -\$2 \quad &\ -\$2 \\ \hline \$1.50x + \$0 &= \$18 \\ \frac{\$1.50x}{\$1.50} &= \frac{\$18}{\$1.50} \\ x &= 12 \end{aligned}$$

Alina rode twelve rides at the fair.

<u>Example</u>

24 feet of fencing was used to enclose a rectangular garden with a width of 8 feet. Write and solve an equation to determine the length of the garden.

The perimeter of a rectangle can be found using the expression $2l + 2w$, where l is the rectangle's length and w is its width. The perimeter of the garden is 24 feet, and the width of the garden is 8 feet, so

$$2l + 2(8) = 24$$
$$2l + 16 = 24$$
$$\underline{-16 \quad -16}$$
$$2l + 0 = 8$$
$$\frac{2l}{2} = \frac{8}{2}$$
$$l = 4$$

The length of the garden is 4 feet.

<u>Example</u>

Leng receives a weekly allowance of five dollars when he completes his usual chores. He can earn an additional fifty cents for each additional chore he does. Write and solve an inequality to find the number of extra chores he should do to earn at least ten dollars this week.

Let n = the number of extra chores Leng must complete. For every extra chore, he earns \$0.50, so $0.5n$ represents the amount of money, in dollars, he will earn from the extra chores. The total amount of money he earns in dollars is represented by the expressions $0.5n + 5$. This weeks, he wishes to earn at least \$10, so he wants to either earn \$10 or more than \$10. Thus, the inequality $5 + 0.5n \geq 10$ represents this scenario.

$$0.5n + 5 \geq 10$$
$$\underline{-5 \quad -5}$$
$$0.5n + 0 \geq 5$$
$$\frac{0.5}{0.5} n \geq \frac{5}{0.5}$$
$$n \geq 10$$

Leng must complete at least 10 extra chores to earn \$10 or more.

Statistics and Probability

Mean, median, and mode

Mean, median, and mode are all measures of central tendency. Measures of central tendency summarize a set of data with values which represent the average, middle, or most common value. The mean, or average, of numerical data can be found by dividing the sum of the numbers in a set by how many numbers are in that set. When numerical data are organized from least to greatest, the middle number or average of the two middle numbers is the median of the set. The mode is the value which appears most frequently in a data set; there may be no mode, one mode, or more than one mode in a set.

<u>Example</u>

Find the mean, median, and mode of these numbers:
23, 42, 36, 21, 28, 29, 32, 28, 40, 36, 39.

The mean of a set of numbers is the sum of the numbers divided by how many numbers are in the set. So, the mean of this set of numbers is $\frac{23+42+36+21+28+29+32+28+40+36+39}{11} = \frac{354}{11} = 32.\overline{18}$.

To find the median, put the numbers in increasing order and find the number in the middle:
21, 23, 28, 28, 29, **32**, 36, 36, 39, 40, 42. The median is 32.

The mode is the value in the set that occurs most frequently. There may be no mode, one mode, or more than one mode in a set. This set has two values that occur twice while all others occur only once. Thus, the modes are those two values, namely 28 and 36.

Random sampling

A random sample is a collection of members, chosen at random and with equal likelihood, from a group about which information is desired. Rather than collecting information from the entire group, which can be quite difficult when the group is very large, information is collected instead from the sample. If the sample is representative of the group and is sufficiently large, then the information gained from the sample is representative can be used to describe the group as a whole.

Example
Determine whether or not each represents a random sample of seventh graders at a middle school:
- A student selects all the seventh graders on her bus.
- A teacher puts the names of her first period students in a hat and draws ten names.
- A principal assigns uses a random number generator to select student ID numbers of seventh graders.

Each seventh grader must be chosen at random and must have an equal chance of being selected. Neither criterion is true in the first scenario. The teacher in the second scenario is selecting a random sample of students in her first period class, but this is not a random sample of the group of seventh graders as a whole. The principle's method of selecting is random and ensures that the likelihood of selecting one student is the same as the likelihood of selecting another.

Example
A random number generator produced the following sets of eight numbers between 1 and 99:
$82, 60, 40, 69, 40, 36, 25, 59$
$98, 28, 15, 91, 51, 74, 11, 36$
$21, 66, 46, 16, 32, 73, 3, 81$
$91, 80, 32, 72, 1, 53, 51, 28.$
Determine the mean of each set. Describe how well the mean of the each random sample represents the set from which the sample was taken.

The mean, or average, is a measure of central tendency which can be found by dividing the sum of the numbers in a set by how many numbers are in that set. The mean of the numbers 1 through 99 is 50, so the mean of a random sample of numbers taken from 1 to 99 should be approximately 50.

$$\frac{82 + 60 + 40 + 69 + 40 + 36 + 25 + 59}{8} = \frac{411}{8} = 51.375$$

$$\frac{98 + 28 + 15 + 91 + 51 + 74 + 11 + 36}{8} = \frac{404}{8} = 50.5$$

$$\frac{21 + 66 + 46 + 16 + 32 + 73 + 3 + 81}{9} = \frac{338}{8} = 42.25$$

$$\frac{91 + 80 + 32 + 72 + 1 + 53 + 51 + 28}{8} = \frac{408}{8} = 51$$

Samples 1, 2, and 4 seem to be a good representatives of the set from which they were taken. The mean of the values in the third sample varies more from the actual mean than the others. One way to improve the likelihood of a random sample's representation of the actual set is to collect more samples.

<u>Example</u>

Suppose all 567 seventh graders in a school vote on which field trip they would like to take from the following options: science center, art museum, or state capitol. Explain how an administrator might use the surveys to predict the most popular vote without tallying all of the results.

An administrator could compile the results from a random sample of surveys. For instance, if the surveys were shuffled, and 50 randomly chosen surveys showed a strong preference for the science center, then the administrator would likely predict that the science center will be the popular vote. To increase her confidence in the results, the administrator could increase the sample size by examining more surveys.

<u>Example</u>

When determining the reading level of a book, a publisher considers many factors, including the average word length. Explain how a publisher might use random sampling to find the average word length in a book.

To find the average word length in a book, a publisher might randomly select a set of words from the book and find the average length of those words. This average should be representative of the whole book if the words are indeed chosen at random.

<u>Example</u>

The histogram below displays the heights of randomly selected eighteen-year-old boys and girls living in Atlanta, Georgia. Compare the variability within each group and between the groups.

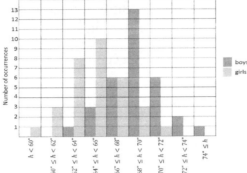

Both of the histograms show a similar, normal distribution of heights, with fewer individuals at the two extremes and the majority clustered at a more central point. The two distributions overlap, which means that some eighteen-year-old boys are the same height as some eighteen-year-old girls. However, there is a noticeable shift to the right in the bell-shaped distributions for boys when compared to girls; this indicates that eighteen-year-old boys are generally taller than girls of the same age. According to the chart, the most common height for eighteen-year-old boys is between 68 and 70 inches, while the most common height among girls is 64 and 66 inches.

<u>Example</u>

Explain what each of the probabilities means in terms of the likelihood of an event:

$$0$$
$$\frac{1}{100,000}$$
$$\frac{1}{2}$$
$$99\%$$
$$1$$

The probability of an event occurring ranges from 0 to 1 when expressed as a fraction or decimal and between 0% and 100% when expressed as a percentage. When there are a finite number of outcomes, a probability of 0 means that the occurrence of an event is impossible, while a probability of 1 means the occurrence is certain. The closer a value is to 0, the less likely it is to occur. For example, a probability of $\frac{1}{100,000}$ indicates that an event is unlikely to occur, whereas a probability of 99% indicates a likely event. Since ½ is halfway between 0 and 1, a probability of ½, or 50%, means that an event is neither likely nor unlikely.

Probability

The set of all outcomes of a probability experiment is called the sample space. For example, if a coin is tossed one time, there are two possible outcomes: heads or tails. So, the sample space consists of two elements. If a coin is tossed two times, there are four possible outcomes: heads then tails, heads then heads, tails then heads, tails then tails. So, the sample space consists of four outcomes.

A simple event consists of only one outcome in the sample space. For instance, the event of getting heads in a single coin toss is a simple event.

Compound events consist of more than one outcome in the sample space. For instance, the event of getting heads at least once in two coin tosses is a compound event. The compound event of getting heads is composed of three outcomes: heads then tails, heads then heads, tails then heads.

Determine the following:
- The probability of winning a single coin toss.
- The probability of a rolling a multiple of 3 on a die.
- The probability of randomly picking a green marble from a bag containing 15 blue marbles and 5 green marbles.
- The probability of randomly picking a red marble from a bag containing 15 blue marbles and 5 green marbles.

Probability is the chance that something will happen. The probability of an event is the ratio of the number of favorable outcomes to the number of possible outcomes when all outcomes are equally likely.

Because there is one favorable outcome of two equally likely outcomes, so the probability of winning a coin toss is ½, or 50%.

Both 3 and 6 are multiples of 3, so there are two favorable outcomes out of six equally likely total outcomes. So, the probability of rolling a multiple of 3 on a die is $\frac{2}{6} = \frac{1}{3} = 33.\overline{3}\%$.

There are five green marbles and fifteen blue marbles in a bag. The probability of picking a green marble is the ratio of green marbles to total marbles, or $\frac{5}{20} = \frac{1}{4} = 25\%$.

Because there are no red marbles in the bag, it is not possible to choose a red marble from the bag. Therefore, the probability of choosing a red marble is 0.

Theoretical probability vs. experimental probability

Theoretical probability is the expected likelihood of an event. Experimental probability is found by conducting trials and comparing the actual occurrence of an event to the number of trials.

For example, the probability of rolling a 2 on a die is 1/6 because there is one favorable outcome, namely rolling a 2, and six equally possible outcomes. So, theoretically, a 2 would appear 100 times if a die is rolled 600 times. Suppose, however, that a die is actually rolled 600 times, and a 2 appears 90 times. The experimental probability is $\frac{90}{600} = \frac{3}{20}$.

If the die is a fair die, the experimental probability should closely approximate or equal the theoretical probability when many trials are conducted.

<u>Example</u>

Determine the probability that the spinner will land on black and predict the number of times the spinner will land on black if the spinner is spun 100 times.

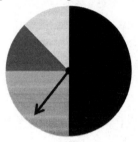

The probabilities of different outcomes on the spinner are not equally likely. Notice that the black section comprises half the circle. So, the spinner will *probably* land in the black section half the time. So, out of 100 spins, the spinner will land in the black section *about* 50 times.

Example

Examine the data collected from 100 coin tosses. Determine the experimental probability of this tossed coin landing on heads (H).

H	H	T	H	T	T	H	T	H	T
T	T	H	T	H	H	T	H	H	T
H	H	T	T	T	T	H	H	T	H
T	H	T	H	H	T	H	H	T	T
H	H	T	H	T	T	H	H	H	T
H	T	T	T	H	H	T	H	T	T
H	T	H	H	T	H	T	T	H	T
H	T	H	H	T	H	T	H	H	T
H	H	T	T	H	T	H	H	T	H
T	T	H	T	H	H	T	T	T	H

Out of 100 trials, 51 coin flips landed on heads, so the experimental probability of getting heads in a coin toss is 51/100. This is very close to the predicted outcome of 50 heads from 100 tosses based on the theoretical probability that a coin will land on heads ($\frac{1}{2}$, or 50%).

Example

The results of 24 rolls of a die are tabulated below. Determine whether or not the results are consistent with the expected results and explain possible reasons for a discrepancy if one exists.

	1	2	3	4	5	6
Number of times rolled	3	4	8	3	2	4

The results of the experiment are not consistent with the expected results. This could be because the die is "loaded" or because enough trials were not performed in the experiment. To determine whether or not the die is loaded, conduct more trials and see if the results are consistent with the expected results based on the experimental probability seen here.

	1	2	3	4	5	6
Number of times rolled	3	4	8	3	2	4
Experimental probability	$\frac{3}{24} = \frac{1}{8}$	$\frac{4}{24} = \frac{1}{6}$	$\frac{8}{24} = \frac{1}{3}$	$\frac{3}{24} = \frac{1}{8}$	$\frac{2}{24} = \frac{1}{12}$	$\frac{4}{24} = \frac{1}{6}$
Theoretical probability	$\frac{1}{6}$	$\frac{1}{6}$	$\frac{1}{6}$	$\frac{1}{6}$	$\frac{1}{6}$	$\frac{1}{6}$
Expected results $\frac{1}{6} \times \frac{24}{1} = \frac{24}{6} = 4$	4	4	4	4	4	4

<u>Example</u>

The results of 24 rolls of a die are tabulated below. Determine the experimental probability from these results and use it to predict the number of times each result will occur if a number cube is rolled 200 times.

	1	2	3	4	5	6
Number of times rolled	3	4	8	3	2	4

The predicted results of 200 rolls based on the experimental probabilities are shown below.

	1	2	3	4	5	6
Number of times rolled	200	200	200	200	200	200
Experimental probability	$\frac{1}{8}$	$\frac{1}{6}$	$\frac{1}{3}$	$\frac{1}{8}$	$\frac{1}{12}$	$\frac{1}{6}$
Predicted results	$\frac{200}{8} = 25$	$\frac{200}{6} \approx 33$	$\frac{200}{3} \approx 67$	$\frac{200}{8} = 25$	$\frac{200}{12} \approx 17$	$\frac{200}{6} \approx 33$

Note that every predicted result is rounded to the nearest whole number since there cannot exist a fraction of a roll. Check to makesure the sum of the rounded numbers is 200: 25+33+67+25+17+33=200.

If the die is indeed loaded, these would be the predicted results. If the die is fair, the results of 200 rolls should show an even distribution of around 33 or 34 rolls over all outcomes.

<u>Example</u>

Joseph is in a math class with 23 other students, 14 of whom are girls. If a student is selected at random from the class, determine the probability that:

- The student selected is Joseph.
- The student selected is a boy.

Since each student has an equally likely chance of being selected from the class, the probability of selecting Joseph is $\frac{1}{number\ of\ students\ in\ the\ class}$. The number of students in the class, including Joseph, is 24. So, $P(Joseph) = \frac{1}{24}$.

Since each student has an equally likely chance of being selected from the class, the probability of selecting a boy is $\frac{number\ of\ boys\ in\ the\ class}{number\ of\ students\ in\ the\ class}$. Since 14 of the 24 students are girls, there are 10 boys. So, $P(boy) = \frac{10}{24} = \frac{5}{12}$.

<u>Example</u>

A baby inherited one copy of a beta-globin gene from her mother and one from her father. Both the baby's mother and father are carriers for sickle cell anemia, meaning that each parent contains a normal allele for beta-globin called type A, and a recessive allele which has a single mutation called type S. The Punnet square below shows the possible genotypes (types of genes) and phenotypes (expressions of the genotypes). Determine the probability that the baby has sickle cell anemia. Express the probabilities as a fraction and as a percent.

	A	S
A	AA (no disease)	AS (no disease/carrier)
S	AS (no disease/carrier)	SS (sickle cell anemia)

The probability that the baby will have sickle cell anemia is ¼, or 25%.

	A	S
A	AA (no disease)	AS (no disease/carrier)
a	AS (no disease/carrier)	SS (sickle cell anemia)

<u>Example</u>

Determine the probability that the baby has inherited a mutated allele from at least one of her parents. Express the probabilities as a fraction and as a percent.

	A	S
A	AA (no disease)	AS (no disease/carrier)
S	AS (no disease/carrier)	SS (sickle cell anemia)

The probability that the baby has inherited a mutated allele from at least one of her parents is ¾, or 75%.

	A	S
A	AA (no disease)	AS (no disease/carrier)
a	AS (no disease/carrier)	SS (sickle cell anemia)

<u>Example</u>

Determine the probability that the baby does not have the disease but carries one copy of the mutated allele. Express the probabilities as a fraction and as a percent.

	A	S
A	AA (no disease)	AS (no disease/carrier)
S	AS (no disease/carrier)	SS (sickle cell anemia)

The probability that the baby does not have the disease but carries one copy of the mutated allele is $\frac{2}{4} = \frac{1}{2}$, or 50%.

	A	S
A	AA (no disease)	AS (no disease/carrier)
a	AS (no disease/carrier)	SS (sickle cell anemia)

<u>Example</u>

Show the sample space for rolling two dice and find the probability of rolling double sixes.

Since there are six outcomes for each die, there are $6 \times 6 = 36$ outcomes for rolling two dice.

1,1	1,2	1,3	1,4	1,5	1,6
2,1	2,2	2,3	2,4	2,5	2,6
3,1	3,2	3,3	3,4	3,5	3,6
4,1	4,2	4,3	4,4	4,5	4,6
5,1	5,2	5,3	5,4	5,5	5,6
6,1	6,2	6,3	6,4	6,5	6,6

For fair dice, each outcome is equally likely. So, the probability can be found $\frac{number\ of\ double\ sixes}{number\ of\ outcomes}$ Only one of the 36 possible outcomes is double sixes, so the probability of getting double sixes is $\frac{1}{36}$.

Use a tree diagram to determine the probability that of two cards pulled from two different decks, at least one is a heart.

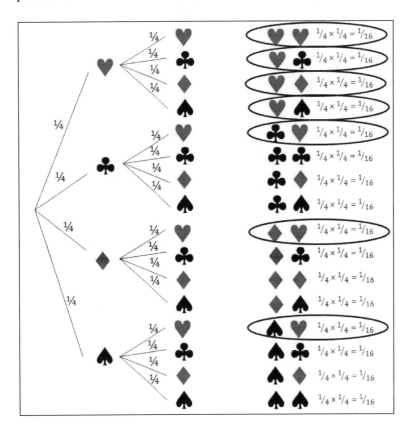

The probability of getting at least one heart includes the outcomes in which a heart is selected from the first deck, in which a heart is selected from the second deck, and in which a heart is selected from both decks. Seven of the sixteen outcomes include at least one heart, so the probability that at least one of the two cards drawn is a heart is $\frac{7}{16}$.

<u>Example</u>

Use a tree diagram to determine the probability that two cards pulled from two different decks are both spades.

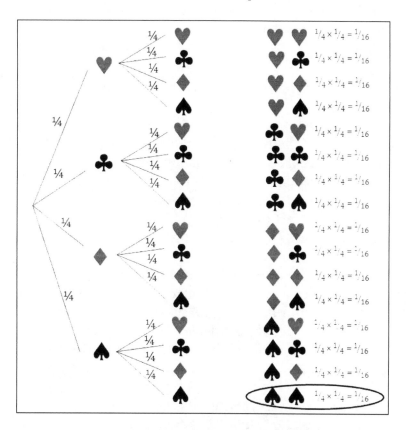

One out of sixteen outcomes is drawing a spade from both decks. So, the probability that both cards are spades is $\frac{1}{16}$.

Notice that the probability of each outcome in the sample space can be found by multiplying the probability of the first event by the probability of the second event.

<u>Example</u>

Suppose a couple has a ¼ chance of having a child with sickle cell anemia. Explain whether or not each statement is true.

- If the couple's first child has sickle cell anemia, then their second child will not have sickle cell anemia.
- If the couple has eight children, two of them will have sickle cell anemia.
- If the couple has six children, it is unlikely that all of them will have sickle cell anemia.

The statement *If the couple's first child has sickle cell anemia, then their second child will not have sickle cell anemia* is **not necessarily true**. If a couple's first child is affected, it is possible for second child to have the disease, too. Each child conceived by the couple has a ¼ chance of developing the disease. The fact that the first child has the disease has no effect on the second child's chance of having it.

The statement *If a couple has eight children, two of them will have sickle cell anemia* is **not necessarily true**. You can <u>predict</u> based on the probability that one in four of the couple's children will be affected that, of eight children, two will have the disease. However, it is possible that none, all, or any number of the children will have the disease.

The statement *If the couple has six children, it is unlikely that all of them will have sickle cell anemia* is **true**. This statement mentions only the relative likelihood of an occurrence. The probability that all six children would have the disease is $\frac{1}{4} \times \frac{1}{4} \times \frac{1}{4} \times \frac{1}{4} \times \frac{1}{4} \times \frac{1}{4} = \frac{1}{4096}$. A probability of $\frac{1}{4096}$ indicates that is unlikely for all six children to have the disease.

Geometry

Example

On a map, the distance between two cities measures 2 ½ inches. A distance of one inch on the map represents an actual distance of 30 miles. Find the actual distance between the two cities.

Use proportional reasoning to find the distance between the two cities. One way to set up a proportion from the given information is to equate two ratios which compare inches to miles, where x represents the unknown distance between the two cities in miles.

$$\frac{2\frac{1}{2}}{x} = \frac{1}{30}$$

There are many ways to solve a proportion. One way is to cross-multiply.

$$2\frac{1}{2} \times 30 = x \times 1$$
$$\frac{5}{2} \times \frac{30}{1} = x$$
$$\frac{150}{2} = x$$
$$75 = x$$

The distance between the cities is 75 miles.

<u>Example</u>

A room has dimensions of 12' wide by 15' long. Using a scale of 1/4":1', draw a blueprint of the room.

First, determine the dimensions of the room on the blueprint. Let w represent the width in inches and l represent the length in inches of the room on the scale drawing.

$$\frac{\frac{1}{4} \text{ in}}{1 \text{ ft}} = \frac{w}{12 \text{ ft}}$$

Using cross-multiplication, $1 \cdot w = \left(\frac{1}{4}\right)(12) = 3$. The width of the room on the blueprint is 3 in.

$$\frac{\frac{1}{4} \text{ in}}{1 \text{ ft}} = \frac{l}{15 \text{ ft}}$$

Again using cross-multiplication, $1 \cdot l = \left(\frac{1}{4}\right)(15) = \frac{15}{4} = 3\frac{3}{4}$. The length of the room on the blueprint is $3\frac{3}{4}$ in.

<u>Example</u>

Below is a box drawn at 1:4 scale.

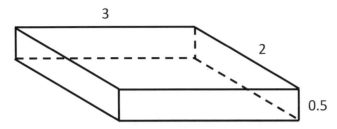

Draw the box at a scale of 1:8.
Determine the actual dimensions of the box.

A scale of 1:4 mean that each side of the box is $\frac{1}{4}$ the size of the actual box.

Redrawing at a scale of 1:8 means that each side of the drawing will be $\frac{1}{8}$ the size of the actual box, or $\frac{1}{2}$ of the size of the given scale drawing since $\frac{1}{2} \times \frac{1}{4} = \frac{1}{8}$. So, draw a box whose dimensions are half of the length, width, and height of the given scale drawing. Since ½=0.5, use either number to calculate the new dimensions. 0.5×3 in $= 1.5$ in 0.5×2 in $= 1$ in 0.5×0.5 in $= 0.25$ in

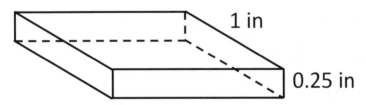

Since each side of the scale drawing is $\frac{1}{4}$ the size of each side of the actual box, the sides of the box are four times the size of the sides in the scale drawing. So, the box's actual dimensions are

- 4×3 in $= 12$ in 4×2 in $= 8$ in 4×0.5 in $= 2$ in.

<u>Example</u>
> The ratio of the lengths of two squares is 1:5.
> - Determine the ratio of the perimeters of the two squares.
> - Determine the ratio of the areas of the two squares.

> If the ratio of the lengths of two squares is 1:5, the ratio of the perimeters of the two squares is also 1:5. Consider, for instance, a square with a length of 1 cm. Its width would also be 1 cm, so it perimeter would be $2l + 2w = 2(1 \text{ cm}) + 2(1 \text{ cm}) = 2 \text{ cm} + 2 \text{ cm} = 4 \text{ cm}$. A square with a length five times that of the first square would have a perimeter of $2(5 \text{ cm}) + 2(5 \text{ cm}) = 10 \text{ cm} + 10 \text{ cm} = 20 \text{ cm}$. So, the ratio of the perimeters of the squares would be 4:20, which reduces to 1:5.

> Considering the same two squares, the area of the first square would be $l \times w = (1 \text{ cm})(1 \text{ cm}) = 1 \text{ cm}^2$. The area of the second would be $(5 \text{ cm})(5 \text{ cm}) = 25 \text{ cm}^2$. So, the ratios of the areas of the squares **would be 1:25.**

<u>Example</u>
> Determine whether or not a triangle can be constructed given
> - The measures of three angles.
> - The lengths of three sides.

> The sum of the measures of a triangle's angles is always 180°. So, a triangle can be constructed from given angle measurements only if those measurements add to 180°.

> The sum of the lengths of two shorter sides of a triangle must be greater than the length of the third side.

<u>Example</u>

Using a ruler and a protractor, draw a right triangle with the two shorter sides measuring 3 in and 4 in. Measure the length of the longest side and the approximate measures of the two non-right angles. Make sure the measurements are consistent with the properties of a triangle.

The length of the longest side is 5 inches. The angle across from the side measuring 4 inches is approximately 53°, and the other angle is approximately 37°. These measures are consistent with the properties of triangles:

- The sum of the angles of the triangle is $53° + 37° + 90° = 180°$. The combined lengths of the two shorter sides exceed the longest side: 3 in + 4 in > 5 *in*. The smallest angle forms the shortest side, and the largest angle forms the longest side.

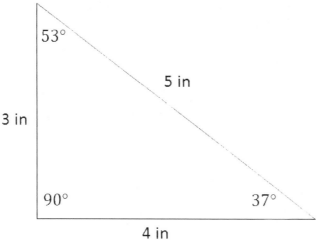

<u>Example</u>

An equilateral triangle whose sides each measure two inches is enlarged by a factor of four. Determine the measurements of the enlarged triangle's sides and angles.

A triangle which has three congruent sides also must have three congruent angles. Since the sum of the angles of a triangle is 180°, each angle measures $\frac{180°}{3} = 60°$. When a triangle is dilated, its angles remain the same; the lengths of the sides change but remain proportionally. Since the scale factor is four, the length of each the three sides is 2 in × 4 = 8 in. So, each angle of the enlarged triangle measures 60°, and each side measures 8 in.

<u>Example</u>

The area, A, of a parallelogram can be found using the formula $A = bh$, where b is the length of the base and h is the height, which is the distance between the parallelogram's base and its opposite, parallel side. Two congruent triangles are obtained when a parallelogram is cut in half from one of its corners to the opposite corner. From this information, determine the area formula of a triangle.

Two identical triangles make up the parallelogram's area, so the area of one triangle is half the area of the parallelogram. Since the area of a parallelogram is $A = bh$, the area of a triangle is $A = \frac{1}{2}bh$, where b is the base of the triangle and h is the height of the triangle as shown.

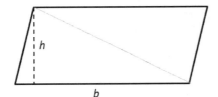 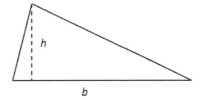

<u>Example</u>

Determine if no triangle, one triangle, or more than one triangle can be drawn given the side and/or angle measurements.
- Side lengths 3 cm, 4 cm, and 5 cm.
- Angle measurements 60°, 40°, and 80°.

In order to construct a triangle, the sum of the lengths of two shorter sides of a triangle must be greater than the length of the third side. $3 \text{ cm} + 4 \text{ cm} > 5 \text{ cm}$, so these sides can be used to draw one triangle.

The sum of the measures of a triangle's angles is always 180°. Since 60°+ 40°=80°, a triangle can be constructed with these angle measurements. In fact, many similar triangles can be constructed with these angle measures.

<u>Example</u>

Determine the two-dimensional cross section obtained by slicing a right, triangular prism parallel to its base

A triangular prism contains two congruent triangular bases which lie in parallel planes. The side edges of a right prism are perpendicular to the base.

Taking the cross section of right, triangular prism parallel to its base yields a triangle which is congruent to the prism's base. One such cross section is illustrated.

<u>Example</u>

Determine the two-dimensional cross section obtained by slicing a right, triangular prism perpendicular to its base

The cross section of right, triangular prism perpendicular to its base yields a rectangle. One such cross section is illustrated.

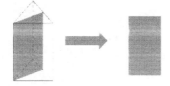

<u>Example</u>

Determine the two-dimensional cross section obtained by slicing a right, rectangular pyramid parallel to its base

A rectangular pyramid is one which has one rectangular base and four triangular sides which meet at an apex. The apex of a right pyramid lies directly above the center of the pyramid's base.

The cross section of a right, rectangular pyramid parallel to its base is a rectangle which is smaller than but similar to the rectangular base. One such cross section is illustrated.

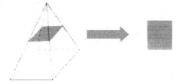

<u>Example</u>

Determine the two-dimensional cross section obtained by slicing a right, rectangular pyramid perpendicular to its base and through its apex.

A rectangular pyramid is one which has one rectangular base and four triangular sides which meet at an apex. The apex of a right pyramid lies directly above the center of the pyramid's base.

The cross section of a right, rectangular pyramid perpendicular to its base and through its apex is a triangle whose height is equal to the height of the pyramid. One such cross section is illustrated.

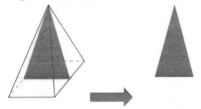

<u>Example</u>

Write the formulas used to find the circumference and area of a circle, respectively.

The formula used to find the circumference, C, of a circle is $C = 2\pi r$ or $C = \pi d$, where r is the radius of the circle and d its diameter. The formula used to find the area, A, of a circle is $A = \pi r^2$, where r is the radius of the circle.

<u>Example</u>

A round table has a diameter of 6 feet. A circular table cloth is cut from a piece of fabric in such a way that it hangs down 6 inches all the way around the table, and a decorative fringe is added along the cut. Using 3.14 as an approximation for π, determine the area of the table cloth and the length of the fringe. Round answers to the nearest tenth.

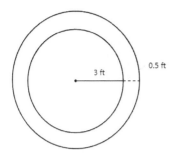

The area of the table cloth is found using the area formula of a circle, $A = \pi r^2$. First, find the radius of the table cloth.

Since the table cloth hangs down 6", which is equivalent to half a foot, add 0.5 ft to the radius of the table to find the radius of the table cloth. The radius of the table cloth is 3.5 ft, so the area of the table cloth is about 38.5 square feet:

$$A = \pi r^2$$
$$A = (3.14)(3.5 \text{ ft})^2$$
$$A = (3.14)(3.5 \text{ ft})^2$$
$$A = (3.14)(12.25 \text{ ft}^2) = 38.465 \text{ ft}^2$$

The fringe goes around the circular table cloth, so its length can be found using the formula for the circumference of a circle, $C = 2\pi r$. The length of the fringe is about 22.0 ft.

$$C = 2\pi r$$
$$C = 2(3.14)(3.5 \text{ ft}) = 21.98 \text{ ft}$$

<u>Example</u>

Name a pair of
Supplementary angles
Complementary angles
Vertical angles
Adjacent angles.

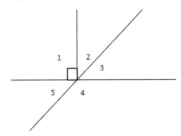

The sum of two supplementary angles is 180°. Angles 4 and 5 are supplementary, as are angles 3 and 4.

The sum of two complementary angles is 90°. Angles 2 and 3 are complimentary.

Vertical angles share a vertex but are not adjacent; rather, vertical angles are congruent angles across from each other in the X made by the intersection of two lines. Angles 3 and 5 are vertical angles.

Adjacent angles share a vertex and a side. Angles 1 and 2, 2 and 3, 3 and 4, 4 and 5, and 1 and 5 are adjacent.

Example

Solve for x and label each angle with its appropriate measure.

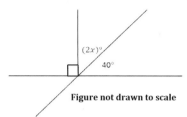

Figure not drawn to scale

Since the angle labeled $(2x)°$ is the complement to the 40° angle, the two add to 90°. So,

$$
\begin{array}{r}
2x \\
+\ 40 \\
=\ 90 \\
\underline{-40} \\
-\ 40 \\
2x + 0 \\
=\ 50 \\
\dfrac{2x}{2} \\
=\ \dfrac{50}{2} \\
x \\
=\ 25
\end{array}
$$

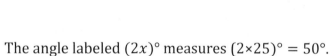

The angle labeled $(2x)°$ measures $(2×25)° = 50°$.

The angle supplementary to the 40° angle must measure 140° since the sum of supplementary angles is 180°.

The angle across from the 40° angle is its vertical angle. Since vertical angles are congruent, that angle also measures 40°.

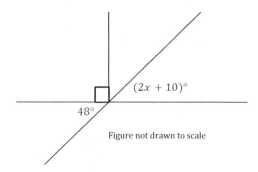

$(2x + 10)°$

48°

Figure not drawn to scale

- 33 -

Since the angle labeled $(2x)°$ is the complement to the 40° angle, the two add to 90°. So,

$$2x$$
$$+ 10$$
$$= 48$$
$$\underline{-10}$$
$$- 10$$
$$2x + 0$$
$$= 38$$
$$\frac{2x}{2}$$
$$= \frac{38}{2}$$
$$x$$
$$= 19$$

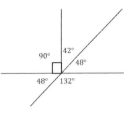

The angles label $(2x + 10)°$ measures 48°, so its complement measures $90° - 48° = 42°$, and its supplement measures $180° - 48° = 132°$.

Example

Match each measurement with the appropriate unit of measurement.
Length of a table
Area of a house
Perimeter of a room
Surface area of a polyhedron

Units
Feet
Square feet
Cubic feet

Volume of a box
Distance between two houses
Storage capacity of a refrigerator

Lengths or distances are measured in units, while area is measured is square units and volume is measured in cubic units. When the unit of measure is feet:

- The length of a table is measured in *feet.*
- The area of a house is measured in *square feet.*
- The perimeter of a room is the distance around the room and is measured in *feet.*
- The surface area of a polyhedron is the sum of the areas of its polygonal faces and is measured in *square feet.*
- The volume of a box is measured in *cubic feet.*
- The distance between two houses is measured in *feet.*
- The storage capacity of a refrigerator is how much space is inside. The refrigerator's volume is measured in *cubic feet.*

<u>Example</u>

Find the surface area and volume of the box.

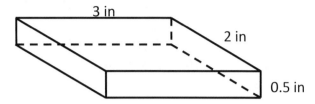

The surface area of a box is the sum of the areas of its six rectangular surfaces.

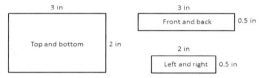

The area of the top rectangle is $A = lw = (3\text{ in})(2\text{ in}) = 6\text{ in}^2$. The area of the top rectangle combined with the bottom rectangle is $6\text{ in}^2 + 6\text{ in}^2 = 12\text{ in}^2$.

The area of the front rectangle is $A = lw = (3\text{ in})(0.5\text{ in}) = 1.5\text{ in}^2$. The area of the front rectangle combined with the back rectangle is $1.5\text{ in}^2 + 1.5\text{ in}^2 = 3\text{ in}^2$.

The area of the left side rectangle is $A = lw = (2\text{ in})(0.5\text{ in}) = 1\text{ in}^2$. The area of the left and right side rectangles together is $1\text{ in}^2 + 1\text{ in}^2 = 2\text{ in}^2$.

The total surface area of the box is $12\ in^2 + 3\ in^2 + 2\ in^2 = 17\ in^2$. The volume, V, of a box is found by $V = lwh$. o, $V = (3\ in)(2in)(0.5\ in) = 3\ in^3$.

<u>Example</u>

An 11'×13' room contains a 3' wide, 7' tall doorway and two 5'x3' windows. The ceiling height is 9'. Determine :

- The price to install baseboards which cost $1.25 per linear foot.
- The price to install flooring which costs $5 per square foot.
- If one gallon of paint which covers 400 square feet of surface is sufficient to paint the walls of the room.

Baseboards run along the edge of the room, but not across doorways. To determine the price for the baseboards, first determine how many feet are needed for the perimeter of the room, excluding the doorway. Use the formula for the perimeter of a rectangle to find the perimeter of the room: $2l + 2w = 2(11 \text{ ft}) + 2(13 \text{ ft}) = 48$ ft. After adjusting for the width of the doorway, 45ft of baseboard is needed for the room. The price of baseboards is 45×1.25=$56.25.

To determine the price for flooring, first determine the area of the room. Use the formula for the area of a rectangle to find the area of the room: $lw = (11 \text{ ft})(13 \text{ ft}) = 143 \text{ ft}^2$. The price for the flooring is 143×5=$715.

To determine the amount of paint needed for the walls, first determine the total surface area to be covered. Two walls are 11'×9' and the other two walls are 13'×9'. Disregarding doors and windows, the total area of the walls is $2(11 \text{ ft})(9 \text{ ft}) + 2(13 \text{ ft})(9 \text{ ft}) = 198 \text{ ft}^2 + 234 \text{ ft}^2 = 432 \text{ ft}^2$. Take away the area of the door, which is $(3 \text{ ft})(7 \text{ ft}) = 21 \text{ ft}^2$, and the area of the windows, which is $2(5 \text{ ft})(3 \text{ ft}) = 30 \text{ ft}^2$: $432 \text{ ft}^2 - 21 \text{ ft}^2 - 30 \text{ ft}^2 = 381 \text{ ft}^2$. The amount of surface which needs paint is 381 ft^2. Purchasing one gallon of paint should be sufficient to cover the walls if only one coat of paint is needed.

Find the perimeter of the shape below.

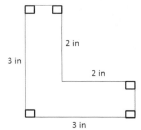

The perimeter of a polygon is the sum of its sides, which is 12 inches.

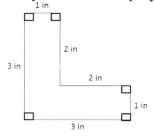

<u>Example</u>

Find the area of the shape below.

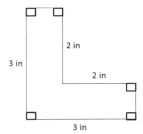

To find the area of an irregular polygon, draw it as a familiar polygon to which another familiar polygon is added or from which a familiar polygon is removed. For example, the shape can be viewed as vertically-oriented rectangle to which a horizontally-oriented rectangle is added, or it can be viewed as a square from which a smaller square is removed.

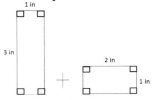

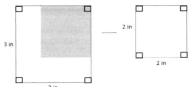

The area of the irregular polygon is the sum of the areas of the two rectangles shown. The area of a rectangle is $A = lw$, so the area of the polygon is $(3\text{ in})(1\text{ in}) + (2\text{ in})(1\text{ in}) = 3\text{ in}^2 + 2\text{ in}^2 = 5\text{ in}^2$.

The area of the irregular polygon is the area of the larger square minus the area of the smaller square. The larger square has side lengths of 3 inches, so the area is (3 in)(3 in)=9 in^2. The smaller square has an area of 4 in^2. So, the area of the irregular polygon is 9 in^2 − 4 in^2 = 5 in^2.

Notice that either method of calculating the irregular polygon's area gives the same answer.

Ratios and Proportional Relationships

Ratio and proportion

A **ratio** is a comparison of two numbers by division. The ratio of a to b, where $b \neq 0$, can be written as
a to b

$$a : b$$

$$\frac{a}{b}$$

A **proportion** is a statement of equality between two ratios. For example, $\frac{a}{b} = \frac{c}{d}$, where $b \neq 0$ and $d \neq 0$, is a proportion equating the ratios $\frac{a}{b}$ and $\frac{c}{d}$.

Percentage

One **Percent** means one part per hundred, so a **percentage** is the ratio of a number to 100. For example, 42% can be written as the ratio $\frac{42}{100}$ or its reduced equivalent, $\frac{21}{50}$.

Example
    Write each percentage as a simplified fraction and as a decimal
- 32%
- 135%

    A percentage is a ratio of a number to 100.
$$32\% = \frac{32}{100} = \frac{8}{25} \ or \ \frac{32}{100} = 0.32$$
$$135\% = \frac{135}{100} = 1\frac{35}{100} = 1\frac{7}{20}$$
$$\frac{135}{100} = 1\frac{35}{100} = 1.35$$

Determining whether or not two ratios form a proportion

Two ratios form a proportion if they are equal. One way to determine if two ratios are equal is to write each ratio as a fraction (as long as neither contains a zero in its denominator) and then **cross-multiply**: that is, multiply the numerator of the first fraction and the denominator of the second; then, multiply the denominator of the first fraction and the numerator of the second. If these two products are equal, the ratios are equal and therefore form a proportion.

$\frac{a}{b} = \frac{c}{d}$ if and only if $a \times d = b \times c$.

For example, $\frac{2}{3} = \frac{12}{18}$, and $2 \times 18 = 3 \times 12 = 36$.

Unit rate

A unit rate is a ratio of two different types of numbers, the second of which is always one. For example, a unit rate can be the number of miles driven in one hour (miles per hour), the price for one ounce of cereal (cents per ounce), or an hourly wage (dollars per hour).

<u>Example</u>

A girl walks half a mile in fifteen minutes. Calculate the unit rate in miles per hour

Since there are sixty minutes in an hour, fifteen minutes is a quarter of an hour: $\frac{15}{60} = \frac{1}{4}$. Since the girl walks ½ mile in ¼ hour, her rate can be written as $\frac{\frac{1}{2}\text{ mile}}{\frac{1}{4}\text{ hour}}$. To determine the unit rate, simplify the fraction so that the denominator is one hour. One way to divide by the fraction $\frac{1}{4}$ is to instead multiply by its reciprocal $\frac{4}{1}$. We get $\frac{1}{2} \div \frac{1}{4} = \frac{1}{2} \times \frac{4}{1} = \frac{4}{2} = 2$. The unit rate is two miles per hour.

<u>Example</u>

A one pound box of cereal costs \$3.20. Calculate the unit price in dollars per ounce

There are sixteen ounces in a pound, so ratio of the price of cereal to its weight can be written as $\frac{\$3.20}{16\text{ oz}}$. To determine the unit price, find the equivalent ratio which compares the price of the cereal to one ounce. To simplify the ratio, divide both the numerator and denominator by 16. Since $\frac{3.20 \div 16}{16 \div 16} = \frac{0.20}{1}$, the unit price of the cereal is twenty cents per ounce.

<u>Example</u>

A bag of 20 cough drops costs $1.68, and a bag of 50 cough drops costs $4.20. Determine which is a better deal

To determine which is the better deal, first find the unit prices of the products. For a bag of 20 cough drops at $1.68, the unit price is $\frac{\$1.68}{20 \text{ cough drops}} = \frac{\$0.084}{1 \text{ cough drop}}$, or 8.4 cents per cough drop. For a bag of 50 cough drops at $4.20, the unit price is $\frac{\$4.20}{50 \text{ cough drops}} = \frac{\$0.084}{1 \text{ cough drop}}$, or 8.4 cents per cough drop. Neither bag is a better bargain than the other since both cost the same amount per cough drop.

<u>Example</u>

Determine what it means for two quantities to have a proportional relationship.

When two quantities have a proportional relationship, there exists a constant of proportionality between the quantities; the product of this constant and one of the quantities is equal to the other quantity. For example, if one lemon costs $0.25, two lemons cost $0.50, and three lemons cost $0.75, there is a proportional relationship between the total cost of lemons and the number of lemons purchased. The constant of proportionality is the unit price, namely $0.25/lemon. Notice that the total price of lemons, t, can be found by multiplying the unit price of lemons, p, and the number of lemons, n: $t = pn$.

Determining whether two quantities have a proportional relationship in a graph

If two quantities are graphed on the coordinate plane, and the result is a straight line through the origin, then the two quantities are proportional.

<u>Example</u>

For the graphs below, determine whether there exists a proportional relationship between x and y. If a proportional relationship exists, find the constant of proportionality and write an equation for the line.

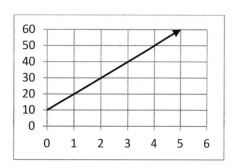

Though the graph of the relationship between x and y is a straight line, it does not pass through the origin. So, though y varies directly as x, the ratio $\frac{y}{x}$ is not constant: for instance, $\frac{20}{1} \neq \frac{30}{2} \neq \frac{40}{3}$..

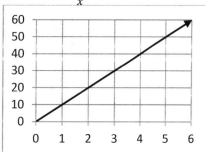

The graph of a proportional relationship is a straight line through the origin.

This graph is a straight line through the origin, so the relationship between x and y is proportional. The constant of proportionality is represented by the ratio $\frac{y}{x}$. This constant is the same as the unit rate. The constant of proportionality is equal to the y-value when $x = 1$. Since the ratio $\frac{y}{x}$ is 10 (see that $\frac{y}{x} = \frac{10}{1} = \frac{20}{2} = \frac{30}{3}$ and so on), or since $y = 10$ when $x = 1$, the constant of proportionality is 10. The relationship between x and y is represented by the equation $y = 10x$.

Determining whether two quantities have a proportional relationship given a table of values

If the ratio of y to x is constant for all values of x and y besides zero, then there is a proportional relationship between the two variables. The value $\frac{y}{x}$ is the constant of proportionality.

Example

Determine whether there exists a proportional relationship between x and y. If a proportional relationship exists, find the constant of proportionality and write an equation to represent the relationship.

x	1	2	3	4
y	5	9	13	17

If the ratio of y to x is constant for all values of x and y besides zero, then there is a proportional relationship between the two variables. The value $\frac{y}{x}$ is the constant of proportionality.

The ratio of y to x is not constant; therefore, the values in the table do not represent a proportional relationship:

x	1	2	3	4
y	5	9	13	17
$\dfrac{y}{x}$	$\dfrac{5}{1} = 5$	$\dfrac{9}{2} = 4.5$	$\dfrac{13}{3} = 4.\overline{3}$	$\dfrac{17}{4} = 4.25$

Example

Determine whether there exists a proportional relationship between x and y. If a proportional relationship exists, find the constant of proportionality and write an equation to represent the relationship.

x	1	2	3	4
y	1	4	9	16

If the ratio of y to x is constant for all values of x and y besides zero, then there is a proportional relationship between the two variables. The value $\frac{y}{x}$ is the constant of proportionality.

The ratio of y to x is not constant; therefore, the values in the table do not represent a proportional relationship:

x	1	2	3	4
y	1	4	9	16
$\dfrac{y}{x}$	$\dfrac{1}{1} = 1$	$\dfrac{4}{2} = 2$	$\dfrac{9}{3} = 3$	$\dfrac{16}{4} = 4$

<u>Example</u>

Determine whether there exists a proportional relationship between x and y. If a proportional relationship exists, find the constant of proportionality and write an equation to represent the relationship.

x	1	2	3	4
y	2.5	5	7.5	10

If the ratio of y to x is constant for all values of x and y besides zero, then there is a proportional relationship between the two variables. The value $\frac{y}{x}$ is the constant of proportionality.

The ratio of y to x is 2.5; therefore, the values in the table represent the proportional relationship modeled by the equation $y = 2.5x$:

x	1	2	3	4
y	2.5	5	7.5	10
$\frac{y}{x}$	$\frac{2.5}{1} = 2.5$	$\frac{5}{2} = 2.5$	$\frac{7.5}{3} = 2.5$	$\frac{10}{4} = 2.5$

<u>Example</u>

Suppose gasoline costs $3 per gallon. Create a table of values for the total cost of gasoline and the gallons of gasoline purchased.

The graph is confined to the first-quadrant of the coordinate plane because neither the amount of gas nor the price of the gas can be negative. Since the cost depends on the number of gallons purchased, plot the number of gallons along the horizontal axis and the cost along the vertical axis.

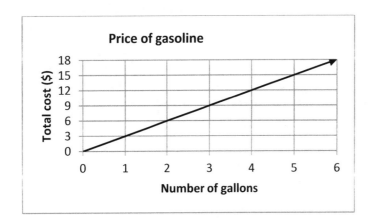

- 44 -

<u>Example</u>

Create a graph of the relationship between the total cost and the gallons of gasoline purchased

The graph is confined to the first-quadrant of the coordinate plane because neither the amount of gas nor the price of the gas can be negative. Since the cost depends on the number of gallons purchased, plot the number of gallons along the horizontal axis and the cost along the vertical axis.

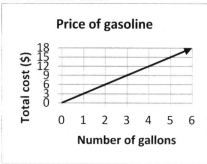

Write an equation which relates total cost to gallons of gasoline purchased. The equation for the line is y=3x, or total cost=3×number of gallons purchased.

<u>Example</u>

Find the rate of travel from a graph representing a proportional relationship between travel time in hours (graphed along the horizontal axis) and distance traveled in miles (graphed along the vertical axis). The slope of the line represents the rate of travel, or the proportionality constant, in miles per hour. The slope of a line is its vertical change, or rise, divided by its horizontal change, or run. These values can be determined from by counting the vertical and horizontal distances between any two points on the line or by using the equation $\frac{y_2-y_1}{x_2-x_1}$, where (x_1, y_1) and (x_2, y_2) are points on the line.

A point on the line shows the distance traveled at a particular travel time. Since the ratio of distance to time is constant along the graph representing a proportion relationship, any point on the graph can be used to find the rate by simply finding the ratio of y to x. For example, a point (3,90) on the graph indicates that it takes three hours to travel ninety miles, so the rate is $\frac{90\ miles}{3\ hours}$= 30 miles per hour.

Since the unit rate, miles per hour, compares the distance traveled in miles to one hour, the unit rate is the y-coordinate when x=1. For example, if the line passes through (1,30), the rate is 30 miles per hour.

- 45 -

Using proportions to solve percent problems

A proportion is a statement of equivalence between two ratios. In percent problems, both ratios compare parts to a whole; in particular, a percentage expresses parts per 100. A proportion which can be used to solved a percent problem is $\frac{part}{whole} = \frac{percent}{100}$.

In the given scenario, 4 is 80% of some number a, so 4 represents part of the unknown number. The proportion, therefore, is $\frac{4}{a} = \frac{80}{100}$. There are many ways to solve proportions. Notice that $\frac{80}{100}$ reduces to $\frac{4}{5}$, so $a = 5$.

<u>Example</u>

A family of six dines at a restaurant which charges an automatic gratuity for parties of six or more. A tip of $28 is added to their bill of $80. Determine the percent gratuity charged.

One way to determine the percent gratuity charged is to set up and solve a proportion of the form

$$\frac{part}{whole} = \frac{percent}{100}.$$

$$\frac{28}{80} = \frac{p}{100}$$

There are many ways to solve proportions. Notice that $\frac{28}{80}$ reduces to $\frac{7}{20}$, which can easily be converted to a fraction with a denominator of 100 by multiplying the numerator and denominator by 5.

$$\frac{7 \times 5}{20 \times 5} = \frac{35}{100}$$

So, $p = 35$. The gratuity added is 35%.

<u>Example</u>

The ratio of flour to sugar in a cookie recipe is 3:1. Find the amount of sugar needed for 1 ½ cups of flour

There are many ways to solve this problem using proportional reasoning. One way is to notice that the amount of flour divided by three gives the amount of sugar.

cups of flour: cups of sugar

3:1

So, the amount of sugar needed for 1 ½ cups of flour can be found by dividing $1\frac{1}{2}$ by 3. $1\frac{1}{2} \div 3 = \frac{3}{2} \times \frac{1}{3} = \frac{3}{6} = \frac{1}{2}$.

$1\frac{1}{2} : \frac{1}{2}$

The amount of sugar needed is ½ cup.

<u>Example</u>

A number is decreased by 20%. The resulting number is then increased by 20%. Determine whether the consequent number is greater than, less than, or equal to the original number

If a number is decreased by 20%, and the resulting number is increased by 20%, then the consequent number will be less than the original number.

Consider, for instance, that the original number is 100.
20% of 100 is $0.20 \times 100 = 20$, and $100 - 20 = 80$.
20% of 80 is $0.20 \times 80 = 16$, and $80 + 16 = 96$.
96 is less than 100.

<u>Example</u>

Suppose you purchase a $7.00 entrée and a $2.00 drink at your favorite restaurant.

- Determine the amount of a 10% tax on your purchase.
- Determine the amount of a 15% tip on the pre-tax amount.
- Find the total price of the meal, including tax and tip.

The amount of your purchase before tax and tip is $9.00. There are many way to calculate the amount of tax on the purchase. One method is to set up and solve a proportion:

The amount of the tax will be calculated as a fraction of the purchase price. That fraction comparing the tax amount to the pre-tax price is equal to 10%, or $\frac{10}{100}$. So, $\frac{tax\ amount}{\$9.00} = \frac{10}{100}$. A tax amount of $0.90 satisfies the proportion.

Another method involves translating the problem into a mathematical expression which represents the tax amount, which is *10% of the purchase*.

A percent is a ratio out of 100, so $10\% = \frac{10}{100} = 0.10$.

The word "of" indicates multiplication.
The purchase price is $9.

So, *10% of the purchase* translates to $0.10 \times \$9$, which equals $0.90.

Again, there are many ways to calculate the amount of the tip. 15% of $9.00 translates to $0.15 \times \$9$, which equals $1.35.

The total price is the cost of the meal plus the tax plus the tip: $9.00+$0.90+$1.35=$11.25.

<u>Example</u>

A school has 400 students; 220 of these students are girls. If the ratio of boys to girls in a class of twenty is representative of the ratio of boys to girls school-wide, determine how many boys are in the class.

Since the ratio of boys to girls in the class is equal to the ratio of boys to girls in the school, the ratio of boys to students in the class must also equal to the ratio of boys to students in the school.

$$\frac{Number\ of\ boys\ in\ the\ school}{number\ of\ students\ in\ the\ school} = \frac{number\ of\ boys\ in\ the\ class}{number\ of\ students\ in\ the\ class}$$

In a school of 400 students, 220 of which are girls, there are $400 - 220 = 180$ boys. The ratio of boys to total students is $180/400$, which reduces to $9/20$. So, in the class of twenty students, there must be nine boys.

Check to see that the ratio of boys to girls in the class is indeed equal to ratio school-wide. If there are nine boys in a class of twenty, then there are eleven girls:

$$\frac{9\ boys}{11\ girls} = \frac{180\ boys}{220\ girls}$$

One way to determine whether or not this statement is true is to cross multiply. If the products are equal, then so are the ratios.

$$9 \times 220 = 1980$$
$$11 \times 180 = 1980$$

The Number System

Integer

The set of **integers** includes whole numbers and their opposites: {...,-3,-2,-1,0,1,2,3...}.

Rational number

A **rational number** is a real number which can be written as a ratio of two integers a and b, where $b \neq 0$ so long as the second is not zero; in other words, any rational number can be expressed in fractional form $\frac{a}{b}$, where $b \neq 0$. Rational numbers include whole numbers, fractions, terminating, and repeating decimals.

<u>Example</u>
> Write each rational number as a fraction
> - 3
> - 0.6
>
> Since dividing any number by one does not change its value, a whole number can be written as a fraction with a denominator of 1. So, $3 = \frac{3}{1}$.
> The six in 0.6 is in the tenths place. The number six-tenths can also be written as $\frac{6}{10}$, which reduces to $\frac{3}{5}$.

<u>Example</u>

Simply each expression

$$\frac{2}{3} + \frac{1}{2}$$

$$\frac{2}{3} - \frac{1}{2}$$

When combining fractions, it is helpful to write them so that they have the same denominator.

1.

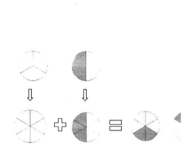

$$\frac{2}{3}$$
$$= \frac{4}{6} \qquad \frac{1}{2}$$
$$= \frac{3}{6}$$

$$\frac{4}{6} + \frac{3}{6}$$
$$= \frac{7}{6}$$
$$= 1\frac{1}{6}$$

2.

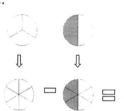

$$\frac{4}{6}$$
$$- \frac{3}{6}$$
$$= \frac{1}{6}$$

Example

Simplify each expression

$$\frac{1}{2} \times \frac{2}{3}$$

$$\frac{1}{8} \div \frac{1}{2}$$

The numerator of the product of two fractions is the product of their numerators; $1 \times 2 = 2$. Likewise, the denominator of the product of two fractions is the product of their denominators: $2 \times 3 = 6$. Reduce the resulting fraction if necessary.

$$\frac{1}{2} \times \frac{2}{3} = \frac{2}{6} = \frac{1}{3}$$

When dividing fractions, rewrite the expression as the product of the first fraction and the reciprocal (or multiplicative inverse) of the second. Reduce if necessary.

$$\frac{1}{8} \div \frac{1}{2} = \frac{1}{8} \times \frac{2}{1} = \frac{2}{8} = \frac{1}{4};$$

Example

Convert each fraction to a decimal.

$$\frac{4}{5}$$

$$\frac{5}{6}$$

A fraction is a quotient of two numbers such that the denominator is not zero. So, one way to convert a fraction to a decimal is to divide the denominator into the numerator. The resulting decimal will either terminate or repeat.

$$\frac{4}{5} = 4 \div 5 = $$

```
    0.8
5 | 4.0
```

$$\frac{5}{6} = 5 \div 6 :$$

```
     0.833...
6 | 5.000...
   0 ↓
   5 0
   4 8↓
     20
     18↓
     20
     18
      2
```

<u>Example</u>

Write each number as a percentage

$$\frac{4}{5}$$

$$\frac{2}{3}$$

$$0.23 = \frac{23}{100} = 23\%$$
$$\frac{4}{5} = \frac{80}{100} = 80\%$$
$$\frac{2}{3} = 0.\bar{6} = 66.\bar{6}\%$$

<u>Example</u>

Express the shaded portion of the circle as a fraction, a decimal, and a percentage.

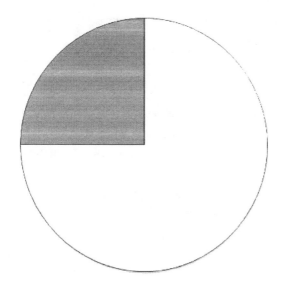

¼ of the circle's area is shaded. $\frac{1}{4} = 0.25 = \frac{25}{100} = 25\%$.

Example

Use a number line to find the sum of 2.1 and 3.2.

Plot 2.1 on a number line and move three and two tenths spaces to the right.

Additive inverse

The sum of a number and its **additive inverse**, or opposite, is the additive identity, 0.

Example

Find the additive inverse of

3

-5

x

The additive inverse of 3 is -3 because $3 + (-3) = 0$.
The additive inverse of -5 is 5 because $-5 + 5 = 0$.
The additive inverse of x is $-x$ because $x + (-x) = 0$.

Multiplicative inverse

The product of a number and its **multiplicative inverse** is the multiplicative identity, 1. The multiplicative inverse of a number is also called its reciprocal. The reciprocal of a non-zero rational number is also rational. Zero does not have a multiplicative because the product of zero and any number is zero and can therefore not equal 1 and because zero can never be in the denominator of a fraction.

Example

Find the multiplicative inverse of

5

$-\dfrac{2}{3}$

x such that $x \neq 0$.

The multiplicative inverse of 5 is $\frac{1}{5}$ because $5\left(\frac{1}{5}\right) = 1$.

The multiplicative inverse of $-\frac{2}{3}$ is $-\frac{3}{2}$ because $-\frac{2}{3}\left(-\frac{3}{2}\right) = 1$.

The multiplicative inverse of x is $\frac{1}{x}$ when $x \neq 0$ because $x\left(\frac{1}{x}\right) = 1$ when $x \neq 0$.

<u>Example</u>

An atom of oxygen has eight positively charged protons and eight negatively charged electrons
- Determine the charge of an atom of oxygen.
- When an atom gains or loses electrons, it becomes an ion. Determine the charge of an oxygen ion which contains two more electrons than an oxygen atom.

An atom of oxygen has a charge of zero because it has the same number of positively charged protons as it does negatively charged electrons: $8 + (-8) = 0$.

An oxygen ion has a charge of -2 because the neutral atom has gained two negatively charged electrons: $0 + (-2) = -2$ or $8 + (-10) = -2$.

<u>Example</u>

Using number lines, show that $2\frac{1}{2} - 2\frac{1}{2} = 2\frac{1}{2} + (-2\frac{1}{2}) = -2\frac{1}{2} + 2\frac{1}{2} = 0$.

To subtract $2\frac{1}{2}$ from $2\frac{1}{2}$ on a number line, start at $2\frac{1}{2}$ and move two and a half spaces to the left.
$$2\frac{1}{2} - 2\frac{1}{2} = 0$$

To simplify $2\frac{1}{2} + (-2\frac{1}{2})$ on a number line, start at $2\frac{1}{2}$ and move two and a half spaces to the left. $2\frac{1}{2} + (-2\frac{1}{2}) = 0$.

Notice that adding -2 ½ to 2 ½ is that same subtracting 2 ½ from 2 ½.

To add $2\frac{1}{2}$ to -2 ½ on a number line, start at -2 ½ and move two and a half spaces to the right. $-2\frac{1}{2} + 2\frac{1}{2} = 0$

As always, the sum of a number and its opposite is zero.
$$2\frac{1}{2} - 2\frac{1}{2} = 2\frac{1}{2} + (-2\frac{1}{2}) = -2\frac{1}{2} + 2\frac{1}{2} = 0$$

<u>Example</u>

Express each as a positive or negative number. Then, write a phrase to represent the number's opposite.

- A gain of four yards
- A deduction of ten points
- A 5°F drop in temperature
- A debit of $1.60
- An extra half-mile

A gain of four yards → **+4** yards. The opposite is *a loss of four yards*, or **−4** yards.

A deduction of ten points → **−10** points. The opposite is *an addition of 10 points*, or **+10** points.

A 5°F drop in temperature → **−5**°F. The opposite is *an increase in temperature of 5 °F*, or **+5**°F.

A debit of $1.60→ **−$1.60**. The opposite is *a credit of $1.60*, or **+$1.60**.

An extra half-mile→ **+½** mile. The opposite is *a half-mile less*, or **−½** mile

Absolute value

The **absolute value** of a number is the number's distance from zero on a number line. A measure of distance is always positive, so absolute value is always positive.

<u>Example</u>

Show that |3|=|−3|.

The absolute value of 3, written as|3|, is 3 because the distance between 0 and 3 on a number line is three units. Likewise, the absolute value of -3, written as |−3|, is 3 because the distance between 0 and -3 on a number line is three units. So, |3|=|−3|.

Multiplying and dividing positive and negative numbers

The product or quotient of two positive numbers is positive.
$$2 \times 4 = 8$$
$$18 \div 3 = 6$$

The product or quotient of two negative numbers is positive.
$$(-3)(-1) = 3$$
$$\frac{-18}{-9} = 2$$

The product or quotient of a positive and a negative number or a negative and a positive number is negative.
$$4(-2) = -8$$
$$-3 \times 6 = -18$$
$$\frac{20}{-10} = -2$$
$$-15 \div 3 = -5$$

<u>Example</u>

For integers p and q, $q \neq 0$, $-\left(\frac{p}{q}\right) = \frac{-p}{q} = \frac{p}{-q}$. Illustrate this property using an example.

Choose an integer value for p and a non-zero integer value for q to show that $-\left(\frac{p}{q}\right) = \frac{-p}{q} = \frac{p}{-q}$. For instance, when p=10 and q=2,

$$-\left(\frac{p}{q}\right) = -\left(\frac{10}{2}\right) = -5$$
$$\frac{-p}{q} = \frac{-10}{2} = -5$$
$$\frac{p}{-q} = \frac{10}{-2} = -$$

<u>Example</u>

A jacket is marked 75% off. Determine which of these methods will give the discounted price of the jacket.
- Find 75% of the jacket's original price and subtract the result from the original price.
- Find 25% of the jacket's original price.
- Divide the original price by four.

All of these methods will give the discounted price of the jacket. If x is the jacket's original price, its new price is $x - 0.75x$. This expression simplifies to $0.25x$. $0.25 = \frac{25}{100} = \frac{1}{4}$, so 0.25x can be rewritten as $\frac{1}{4}x$, which equals $\frac{x}{4}$.

Example

A person plans to lose ½ pound each week by following a healthy diet. Write and simplify an expression to show his expected weight loss after six weeks of healthy eating

A loss of ½ pound each week for six weeks translates to $\left(-\frac{1}{2}\right)(6)$, which simplifies to -3. He can expect to lose three pounds in six weeks.

Example

A gymnast's routine has a start value of 16 points. During her routine, she incurs thee deductions of one-tenth of a point, two deductions of three-tenths of a point, and one deduction of half a point. Determine the score she receives for her performance.

Write an expression to represent the gymnast's score. Each deduction is subtracted from her start value. Write all the deductions as fractions or as decimals. $\frac{1}{10} = 0.1; \frac{3}{10} = 0.3; \frac{1}{2} = 0.5$.
$$16 - 3(0.1) - 2(0.3) - 0.5$$
$$16 - 0.3 - 0.6 - 0.5$$
$$15.7 - 0.6 - 0.5$$
$$15.1 - 0.5 = 14.6$$

The gymnast's score is 14.6.

Example

A service provider charges $25.75 for phone, $27.75 for internet, and $33.50 for cable each month. A one-time credit of $35.50 is applied towards a customer's bill when the customer opts to prepay for service by quarterly bank draft. After a new customer orders phone, internet, and cable service and signs up for automatic bill pay, she notices a transaction on her bank statement of -$225.50. Write an expression which justifies this charge made by the service provider.

If the customer pays her bill quarterly, then she pays for three months of service at one time. So, her bill includes three times the total for the phone charge and the internet charge and the cable charge. A credit of $35.00 is given only once.
$$3[(-\$25.75) + (-\$27.75) + (-\$33.50)] + \$35.00$$
$$= 3(-\$87.00) + \$35.50$$
$$= -\$261.00 + \$35.50$$
$$= -\$225.50$$

Practice Test #1

Practice Questions

1. Ana has completed approximately $\frac{2}{7}$ of her research paper. Which of the following best represents the percentage of the paper she has completed?

    Ⓐ 24%

    Ⓑ 26%

    Ⓒ 27%

    Ⓓ 29%

2. Simplify the expression: $2n + (3n - 2)^2$

---

3. Elijah has prepared $2\frac{1}{2}$ gallons of lemonade to distribute to guests at a party. If there are 25 guests, how much lemonade is available to each guest, given that each guest receives an equal amount?

    Ⓐ $\frac{1}{8}$ of a gallon

    Ⓑ $\frac{1}{6}$ of a gallon

    Ⓒ $\frac{1}{12}$ of a gallon

    Ⓓ $\frac{1}{10}$ of a gallon

4. The points M, N, and O are plotted on the number line below. Plot point P based on the equation: $N - M + O = P$

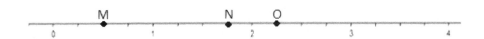

5. Part A: Edward spins the spinner below three times. If the spinner lands on a different number each time what is the highest total he could get?

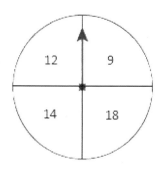

_____

Part B: He decides to spin it one more time. What is the probability that he will land on a number that he has already landed on?

_____

6. A bag of coffee costs $9.85 and contains 16 ounces of coffee. Which of the following best represents the cost per ounce?

Ⓐ $0.67

Ⓑ $0.64

Ⓒ $0.65

Ⓓ $0.62

7. Which of the following is equivalent to $4^3 + 12 \div 4 + 8^2 \times 3$?

Ⓐ 249

Ⓑ 393

Ⓒ 211

Ⓓ 259

8. Part A: The ingredients needed for a cake are given below:

2 eggs

$1\frac{3}{4}$ cups of flour

2 teaspoons baking soda

½ cup butter

$1\frac{1}{4}$ teaspoons vanilla extract

$\frac{3}{4}$ cup of milk

What is the ratio of butter to flour?

    Ⓐ 1:3

    Ⓑ 2:5

    Ⓒ 2:7

    Ⓓ 3:2

Part B: When Lucy is making the cake she accidently puts a whole cup of milk in it. What was the ratio of milk to flour before and what is it now?

_____   _____

9. Part A: The original price of a jacket is $36.95. The jacket is discounted by 25%. Before tax, which of the following best represents the cost of the jacket?

    Ⓐ $27.34

    Ⓑ $27.71

    Ⓒ $28.82

    Ⓓ $29.56

Part B: If tax is 8% how much does it cost?

_____

10. Martin and his friends are taking a road trip from Houston, TX to Las Vegas, NV. The trip will take two days and they will spend the night in El Paso, TX. The first day they drove for 11 hours at a rate of 69 miles per hour. The next day they drove for 10 hours at a rate of 72 miles per hour. How far is it from Houston to Las Vegas?

_____

11. A bottle of lotion contains 20 fluid ounces and costs $3.96. Which of the following best represents the cost per fluid ounce?

Ⓐ $0.18

Ⓑ $0.20

Ⓒ $0.22

Ⓓ $0.24

12. Solve the equation for x: $3^2 + 2x = 17$.

_____

13. Given the figure below what is the area of the shaded regions? Figure is not to scale.

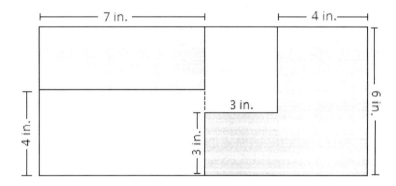

_____

14. Given the sequence represented in the table below, where n represents the position of the term and a_n represents the value of the term, which of the following describes the relationship between the position number and the value of the term?

n	1	2	3	4	5	6
a_n	5	2	−1	−4	−7	−10

Ⓐ Multiply n by 2 and subtract 4

Ⓑ Multiply n by 2 and subtract 3

Ⓒ Multiply n by −3 and add 8

Ⓓ Multiply n by −4 and add 1

15. The number 123 is the 11ᵗʰ term in a sequence with a constant rate of change. Which of the following sequences has this number as its 11ᵗʰ term?

Ⓐ 5, 17, 29, 41, ...

Ⓑ 3, 15, 27, 39, ...

Ⓒ −1, 11, 23, 35, ...

Ⓓ 1, 13, 25, 37, ...

16. Kevin pays $12.95 for a text messaging service plus $0.07 for each text message he sends. Which of the following equations could be used to represent the total cost, y, when x represents the number of text messages sent?

Ⓐ $y = \$12.95x + \$0.07$

Ⓑ $y = \$13.02x$

Ⓒ $y = \dfrac{\$12.95}{\$0.07}x$

Ⓓ $y = \$0.07x + \$12.95$

17. Hannah draws two supplementary angles. One angle measures 34°. What is the measure of the other angle?

Ⓐ 56°

Ⓑ 66°

Ⓒ 146°

Ⓓ 326°

18. Part A: Steven's class had a pushup contest and the results are recorded below.

What is the median number of pushups the class did?

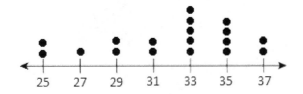

_____

Part B: What is the difference between the median and the mean number of pushups?

_____

19. A triangle has the following angle measures: 98°, 47°, and 35°. What type of triangle is it?

Ⓐ Obtuse

Ⓑ Right

Ⓒ Acute

Ⓓ Equiangular

20. Part A: Jordan has a bag full of red and blue marbles. There are 16 red marbles and 24 blue marbles? What is the probability of him drawing a red marble?

_____

Part B: Jordan has continued to draw marbles and has now taken out 4 red and 6 blue. What is the probability of him drawing a red marble now?

_____

21. **Which figure has two circular bases and a lateral face?**

Ⓐ Cone

Ⓑ Prism

Ⓒ Cylinder

Ⓓ Sphere

22. **Plot the two ordered pairs on the graph below: (x-3, 5) and (7-x, -2) where x = 2.**

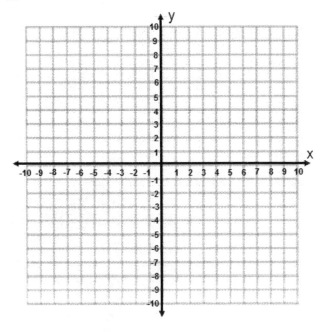

23. **Bookstore A sells a particular book for $15.25, but they have it on sale for 20% off. Bookstore B sells it for $12.45. How much more does it cost at Bookstore B?**

_____

24. **A carpenter must fix a broken section of a kitchen cabinet. The intact portion of the cabinet forms a 76 degree angle with the wall. The width of the cabinet is supposed to form a 90 degree angle with the wall. What angle measure should the carpenter use when cutting the piece that will fit next to the 76 degree angle?**

Ⓐ 12°

Ⓑ 14°

Ⓒ 104°

Ⓓ 136°

25. Which of the following represents the net of a triangular prism?

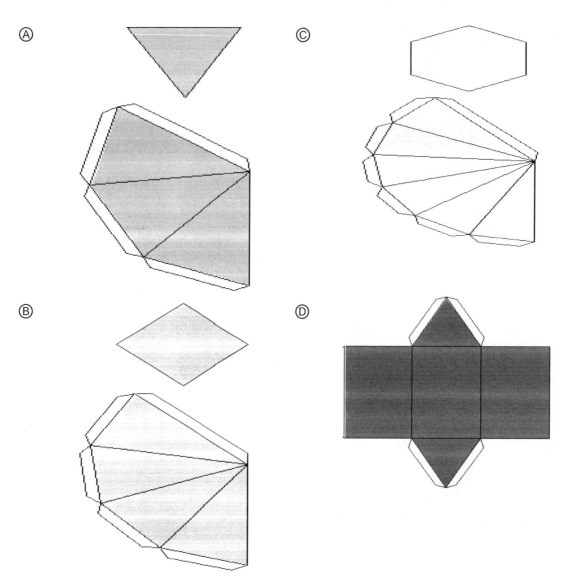

Ⓐ

Ⓒ

Ⓑ

Ⓓ

26. A circle has a radius of 23 cm. Which of the following is the best estimate for the circumference of the circle?

Ⓐ 71.76 cm

Ⓑ 143.52 cm

Ⓒ 144.44 cm

Ⓓ 72.22 cm

27. Sally is driving to the store. It takes her $\frac{1}{12}$ of an hour to go 3.4 miles. How many miles an hour is Sally driving?

Ⓐ 36 mph

Ⓑ 17 mph

Ⓒ 40.8 mph

Ⓓ 42.4 mph

28. Which of the following is also equal to $4n^2 + (3n + 5)^2$? Select all that apply.

I. $2n^2 + 2n^2 + (3n + 5)^2$
II. $4n^2 + (3n^2 + 25)$
III. $4n^2 + (3n + 5)$
IV. $13n^2 + 30n + 25$
V. $7n^2 + 25$

29. Ashton draws the parallelogram shown below. How many square units represent the area of the parallelogram?

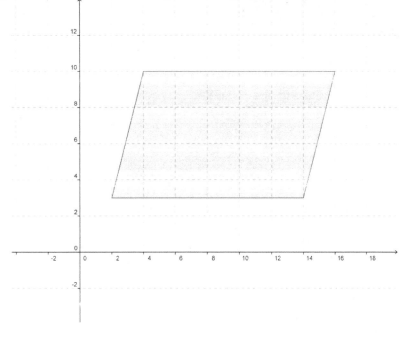

30. In the formula for the volume of the figure shown below, written as $V = B \cdot h$, h represents the height of the prism when it rests one of its bases. What does the B represent?

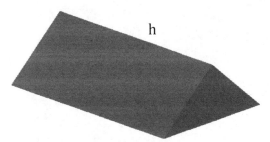

h

Ⓐ $\frac{1}{3}bh$, where b represents the length of the triangle's base and h represents the triangle's height

Ⓑ bh, where b represents the length of the triangle's base and h represents the triangle's height

Ⓒ $2bh$, where b represents the length of triangle's base and h represents the triangle's height

Ⓓ $\frac{1}{2}bh$, where b represents the length of triangle's base and h represents the triangle's height

31. A rectangular prism has a length of 14.3 cm, a width of 8.9 cm, and a height of 11.7 cm. Which of the following is the best estimate for the volume of the rectangular prism?

Ⓐ $1{,}512 \text{ cm}^3$

Ⓑ $1{,}287 \text{ cm}^3$

Ⓒ $1{,}386 \text{ cm}^3$

Ⓓ $1{,}620 \text{ cm}^3$

32. A can has a radius of 3.5 cm and a height of 8 cm. Which of the following best represents the volume of the can?

Ⓐ 294.86 cm^3

Ⓑ 298.48 cm^3

Ⓒ 307.72 cm^3

Ⓓ 309.24 cm^3

33. Fred designs a candy box in the shape of a triangular prism. The base of each triangular face measures 4 inches, while the height of the prism is 7 inches. Given that the length of the prism is 11 inches, what is the volume of the candy box?

Ⓐ 102 in^3

Ⓑ 128 in^3

Ⓒ 154 in^3

Ⓓ 308 in^3

34. Miranda rolls a standard die and spins a spinner with 4 equal sections. Which of the following represents the sample space?

Ⓐ 10

Ⓑ 12

Ⓒ 24

Ⓓ 36

35. A hat contains 6 red die, 4 green die, and 2 blue die. What is the probability that Sarah pulls out a blue die, replaces it, and then pulls out a green die?

Ⓐ $\frac{1}{18}$

Ⓑ $\frac{1}{3}$

Ⓒ $\frac{1}{2}$

Ⓓ $\frac{1}{16}$

36. The histogram below represents the overall GRE scores for a sample of college students. Which of the following is a true statement?

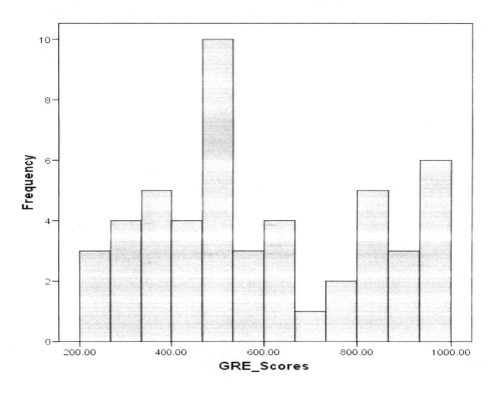

Ⓐ The range of GRE scores is approximately 600

Ⓑ The average GRE score is 750

Ⓒ The median GRE score is approximately 500

Ⓓ The fewest number of college students had an approximate score of 800

37. What is the area of the circle on the graph below? Each square represents 1 inch.

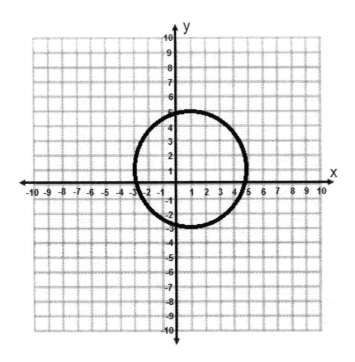

---

38. Raymond has the triangular prism shown below. If he cuts two-dimensional slices out of it how many different shapes could he make?

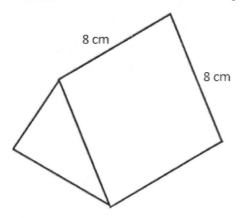

Ⓐ 2, a triangle and a rectangle

Ⓑ 1, a triangle

Ⓒ 2, a square and a rectangle

Ⓓ 3, a square, a triangle, and a rectangle`

39. Gabriel went to a taco shop for lunch. He ordered 3 tacos and paid $4.14. The next day he decided to try a different taco shop and ordered 4 tacos and paid $4.80. How much more expensive was the first taco shop per taco? Express your answer as a percent.

_____

40. Amy rolled a die and flipped a coin and spun a spinner with four equal sections numbered 1-4. What is the probability that she rolled an even number, got heads and then spun an even number?

Ⓐ $\frac{1}{4}$

Ⓑ $\frac{1}{2}$

Ⓒ $\frac{3}{4}$

Ⓓ $\frac{1}{8}$

Answers and Explanations

1. D: In order to convert the given fraction to a percentage, divide 2 by 7. Doing so gives a decimal of approximately 0.29. The decimal can be converted to a percentage by multiplying by 100, which moves the decimal point two places to the right and gives 29%.

2. $2n + (3n - 2)^2 = 2n + 9n^2 - 6n - 6n + 4 = 9n^2 - 10n + 4$

3. D: In order to determine the amount available to each guest, the total amount of prepared lemonade should be divided by 25 guests. Thus, the expression $2\frac{1}{2} \div 25$ represents the amount that each guest has available for consumption. The mixed fraction can be rewritten as $\frac{5}{2}$. The expression can be simplified by writing $\frac{5}{2} \div 25 = \frac{5}{2} \times \frac{1}{25}$, which equals $\frac{5}{50}$, or $\frac{1}{10}$.

4. M=$\frac{1}{2}$, N=$1\frac{3}{4}$, and O=$2\frac{1}{4}$, so N-M+O=$3\frac{1}{2}$. The number line is shown below.

5. Part A: 44: I the spinner must land on a different number each time then the highest three numbers are 12,14, and 18. Added together they equal 44.

Part B: $\frac{3}{4}$: He has already landed on three of the four numbers so the probability is $\frac{3}{4}$.

6. D: The cost per ounce can be calculated by dividing the cost of the bag by the number of ounces the bag contains. Thus, the cost per ounce can be calculated by writing $9.85 ÷ 16, which equals approximately $0.62 per ounce.

7. D: The order of operations states that numbers with exponents must be evaluated first. Thus, the expression can be rewritten as $64 + 12 \div 4 + 64 \times 3$. Next, multiplication and division must be computed as they appear from left to right in the expression. Thus, the expression can be further simplified as $64 + 3 + 192$, which equals 259.

8. Part A: C: Both flour and butter are given in cups so they are easy to compare. First find a like denominator. $\frac{1}{2}$ is equal to $\frac{2}{4}$, and $1\frac{3}{4}$ is equal to $\frac{7}{4}$. So the ratio is 2:7.

Part B: 3:7, 4:7: This question is very similar to Part A. Both milk and flour are given in cups and fourths. It's $\frac{3}{4}$ to $\frac{7}{4}$ or 3:7, and if she puts a whole cup in then its $\frac{4}{4}$ to $\frac{7}{4}$ or 4:7.

9. Part A: B: The discounted price is 25% less than the original price. Therefore, the discounted price can be written as $36.95 - ((0.25)(36.95))$, which equals approximately 27.71. Thus, the discounted price of the jacket is $27.71.

Part B: $29.93: If sales tax is 8% then that can be written as ($27.71)(1.08), which is approximately $29.93.

10. 1479 miles: The first day they went 69 miles per hour for 11 hours, so they went $11 \times 69 = 759 \ miles$. The second day they went 72 miles per hour for 10 hours, so they went $10 \times 72 = 720 \ miles$. Total they went 1479 miles.

11. B: In order to find the unit rate, the cost of the bottle should be divided by the number of fluid ounces contained in the bottle: $\frac{\$3.96}{20} \approx 0.20$. Thus, the cost per fluid ounce is approximately $0.20.

12. X=4: $3^2 + 2x = 17, 9 + 2x = 17, 2x = 8, x = 4$

13. 45 square inches: The top left shaded region can be found by first finding the width. Since 6 in. is given as the width of the whole rectangle and 4 in. is given for the width of the non shaded region then the width of the shaded region is the difference of 2 inches. So, the area of that region is $7 \ in. \times 2in. = 14 \ square \ inches$. The other shaded region can be broken into a 3 in. by 3 in. square and a 4 in. by 6 in. rectangle. So, $3 \ in. \times 3 \ in. = 9 \ ssquare \ inches$ and $4 \ in. \times 6 \ in. = 24 \ square \ inches$. Added together the total area is 45 square inches.

14. C: The equation that represents the relationship between the position number, n, and the value of the term, a_n, is $a_n = -3n + 8$. Notice each n is multiplied by –3, with 8 added to that value. Substituting position number 1 for n gives $a_n = -3(1) + 8$, which equals 5. Substitution of the remaining position numbers does not provide a counterexample to this procedure.

15. B: All given sequences have a constant difference of 12. Subtraction of 12 from the starting term, given for Choice B, gives a y-intercept of –9. The equation $123 = 12x - 9$ can thus be written. Solving for x gives $x = 11$; therefore, 123 is indeed the 11th term of this sequence. Manual computation of the 11th term by adding the constant difference of 12 also reveals 123 as the value of the 11th term of this sequence.

16. D: The constant amount Kevin pays is $12.95; this amount represents the y-intercept. The variable amount is represented by the expression $0.07x$, where x represents the number of text messages sent and $0.07 represents the constant rate of change or slope. Thus, his total cost can be represented by the equation $y = \$0.07x + \$12.95$.

17. C: Supplementary angles add to 180 degrees. Therefore, the other angle is equal to the difference between 180 degrees and 34 degrees: $180 - 34 = 146$. Thus, the other angle measures $146°$.

18. Part A: 33: The median number is the middle number out of the group. In this case there are 18 numbers so it is the average of the 9th and 10th number. However, since the 9th and 10th numbers are both 33 it is just 33.

Part B: 1: The mean can be found by adding all of the numbers together and dividing by 18. If you add all of the numbers up and divide by 18 you get 32. The difference in the mean and median is 1.

19. A: A triangle with an obtuse angle (an angle greater than $90°$) is called an obtuse triangle.

20. Part A: $\frac{2}{5}$: There are a total of 40 marbles in the bag. The probability of him drawing a red one is $\frac{16}{40}$ or $\frac{2}{5}$.

Part B: $\frac{2}{3}$: If 4 red and 6 blue are missing then there are only 12 red and 18 blue remaining. The probability of drawing a red one then is $\frac{12}{18}$ or $\frac{2}{3}$.

21. C: A cylinder has two circular bases and a rectangular lateral face.

22. The point (x-3, 5) is (2-3, 5) or (-1, 5). The point (7-x, -2) is (7-2, -2) or (5, -2). Both are graphed below.

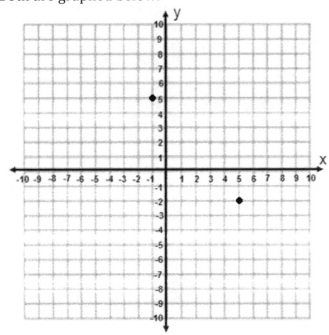

23. $.25: First find the cost of the book at Bookstore A. The price would be $15.25(.8) = $12.20. The cost at Bookstore B is $12.45 so it is $.25 more.

24. B: Since the intact portion of the cabinet and the missing piece form a 90 degree angle with the wall, the missing piece must have an angle equal to the difference between 90 degrees and 76 degrees. Thus, the newly cut cabinet piece should have an angle measure of 14 degrees.

25. D: A triangular prism has two triangular bases and three rectangular faces.

26. C: The circumference of a circle can be determined by using the formula $C = \pi d$. A radius of 23 cm indicates a diameter of 46 cm, or twice that length. Substitution of 46 cm for d and 3.14 for π gives the following: $C = 3.14 \cdot 46$, which equals 144.44. Thus, the circumference of the circle is approximately 144.44 cm.

27. C: If she goes 3.4 miles in $\frac{1}{12}$ of an hour then just multiply by 12 to see how far she will go in one hour. Then that is her miles per hour.

28. I, IV: Start by simplifying the equation. $4n^2 + (3n + 5)^2 = 4n^2 + 9n^2 + 15n + 15n + 25 = 13n^2 + 30n + 25$. So, you can see that you came up with answer IV right there, and answer I is the same as the original equation except for the $4n^2$ is broken down to $2n^2 + 2n^2$.

29. 84: The area of a parallelogram can be found by using the formula $A = bh$, where b represents the length of the base and h represents the height of the parallelogram. The base and the height of the parallelogram are 12 units and 7 units, respectively. Therefore, the area can be written as $A = 12 \cdot 7$, which equals 84.

30. D: The B in the formula $V = Bh$ represents the area of the triangular base. The formula for the area of a triangle is $\frac{1}{2}bh$, where b represents the length of the triangle's base and h represents the triangle's height.

31. A: The dimensions of the rectangular prism can be rounded to 14 cm, 9 cm, and 12 cm. The volume of a rectangular prism can be determined by finding the product of the length, width, and height. Therefore, the volume is approximately equal to $14 \times 9 \times 12$, or 1,512 cm^3.

32. C: The volume of a cylindrical can be found using the formula $V = \pi r^2 h$, where r represents the radius and h represents the height. Substitution of the given radius and height gives $V = \pi (3.5)^2 \cdot 8$, which is approximately 307.72. Thus, the volume of the can is approximately 307.72 cm^3.

33. C: The volume of a triangular prism can be determined using the formula $V = \frac{1}{2}bhl$, where b represents the length of the base of each triangular face, h represents the height of each triangular face, and l represents the length of the prism. Substitution of the given values into the formula gives $V = \frac{1}{2} \cdot 4 \cdot 7 \cdot 11$, which equals 154. Thus, the volume of the candy box is 154 cubic inches.

34. C: The sample space of independent events is equal to the product of the sample space of each event. The sample space of rolling a die is 6; the sample space of spinning a spinner with four equal sections is 4. Therefore, the overall sample space is equal to 6 × 4, or 24.

35. A: The events are independent since Sarah replaces the first die. The probability of two independent events can be found using the formula $P(A \text{ and } B) = P(A) \cdot P(B)$. The probability of pulling out a blue die is $\frac{2}{12}$. The probability of pulling out a green die is $\frac{4}{12}$. The probability of pulling out a blue die and a green die is $\frac{2}{12} \cdot \frac{4}{12}$, which simplifies to $\frac{1}{18}$.

36. C: The score that has approximately 50% above and 50% below is approximately 500 (517 to be exact). The scores can be manually written by choosing either the lower or upper end of each interval and using the frequency to determine the number of times to record each score, i.e., using the lower end of each interval shows an approximate value of 465 for the median; using the upper end of each interval shows an approximate value of 530 for the median. A score of 500 (and the exact median of 517) is found between 465 and 530.

37. 16π or 50.27 inches: The area of a circle is πr^2. You can find the radius by counting the number of units across the circle and dividing by 2. So the radius is 4 inches. $\pi 4^2 = 16\pi \text{ or } 50.27 \text{ inches}$.

38. D: If you were to cut a slice vertically it would produce a triangle. If you were to cut a slice long ways it would produce a rectangle, and if you were to cut a slice that was the front face of the prism it would be an 8 by 8 square.

39. 15%: First find the cost of the first taco shop by dividing $4.14 by 3, which comes out to $1.38 per taco. Then find the price at the second shop by dividing $4.80 by 4 to get $1.20 per taco. The difference is .$18. To get a percentage divide $.18 by $1.20 and multiply by 100. The first shop cost 15% more than the second.

40. D: The probability of getting an even number is $\frac{3}{6}$. The probability of getting heads is $\frac{1}{2}$. The probability of spinning an even number is $\frac{2}{4}$. The probability of all three occurring can be calculated by multiplying the probabilities of the individual events: $\frac{3}{6} \cdot \frac{1}{2} \cdot \frac{2}{4}$ equals $\frac{1}{8}$.

Practice Test #2

Practice Questions

1. Which of the following is the largest number?

$\frac{14}{4}, 3.41, \pi, 3\frac{3}{8}$

_____

2. A plane takes off from Dallas and lands in New York 3 hours and 20 minutes later. The distance from Dallas to New York is 1510 miles. Approximately how fast was the plane traveling?

Ⓐ 445 mph

Ⓑ 453 mph

Ⓒ 456 mph

Ⓓ 449 mph

3. According to the order of operations, which of the following steps should be completed immediately following the evaluation of the squared number when evaluating the expression $9 - 18^2 \times 2 + 12 \div 4$?

Ⓐ Subtract 18^2 from 9

Ⓑ Multiply the squared value by 2

Ⓒ Divide 12 by 4

Ⓓ Add 2 and 12

4. A parcel of land has 35 mature trees for every 3 acres. How many mature trees can be found on 18 of the acres?

Ⓐ 206

Ⓑ 212

Ⓒ 210

Ⓓ 214

5. Which of the following is equivalent to $-8^2 + (17 - 9) \times 4 + 7$?

Ⓐ −217

Ⓑ 24

Ⓒ −64

Ⓓ −25

6. Jason chooses a number that is the square root of four less than two times Amy's number. If Amy's number is 20, what is Jason's number?

Ⓐ 6

Ⓑ 7

Ⓒ 8

Ⓓ 9

7. Part A: What is the area of the shaded region below?

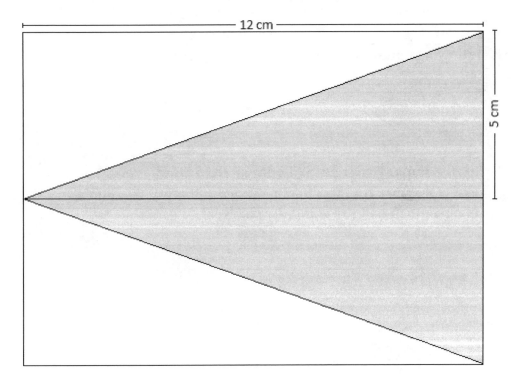

Part B: How does the shaded region compare to the non shaded region?

Ⓐ The shaded region is bigger than the non shaded region

Ⓑ Both the shaded and non shaded region are the same size

Ⓒ The non shaded region is bigger than the shaded region

Ⓓ The area of the non shaded region cannot be determined

8. Given the table below what would *y* be if *x=5*?

X	-2	0	3	4
y	2	-2	7	14

Ⓐ 21

Ⓑ 23

Ⓒ 19

Ⓓ 24

9. Given the following equation what is *x* equal to if *y* equals 8? $6x + 4 = 2y - 7$.

_____

10. A landscaping company charges \$25 per $\frac{1}{2}$-acre to mow a yard. The company is offering a 20% discount for the month of May. If Douglas has a two-acre yard, how much will the company charge?

Ⓐ \$65

Ⓑ \$80

Ⓒ \$70

Ⓓ \$75

11. A house is priced at $278,000. The price of the house has been reduced by $12,600. Which of the following best represents the percentage of the reduction?

Ⓐ 3%

Ⓑ 4%

Ⓒ 5%

Ⓓ 6%

12. Amy buys 4 apples and 3 bananas at the grocery store. She spent a total of $5.17. Each apple cost $.94. If each banana cost the same amount how much did one banana cost?

_____

13. Melanie makes $12 an hour and is taxed at 15% on her income. Lynn makes $14 an hour and is taxed at 18% on her income. If they both work a 40 hour week, how much more does Lynn make than Melanie?

_____

14. A cone has a radius of 4 cm and an approximate volume of 150.72 cm³. What is the height of the cone?

Ⓐ 7 cm

Ⓑ 9 cm

Ⓒ 8 cm

Ⓓ 12 cm

15. What is the range of the points on the number line below?

_____

16. Complete the equation below.

$$2\frac{5}{8} - \frac{3}{2} + \left(\frac{2}{3} - \frac{1}{6}\right) = 1\frac{1}{8} + \underline{\hspace{2cm}} = \underline{\hspace{2cm}}$$

17. Point S and Point T are shown on the number line below. Which of the following equations produces a point, R, not on the number line?

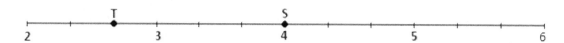

Ⓐ 2S-T=R

Ⓑ 2T-S=R

Ⓒ 3T-S=R

Ⓓ 3S-2T=R

18. Part A: The spinner below is spun once. What is the probability of landing on a 12 or greater?

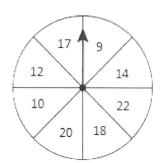

Ⓐ $\frac{5}{8}$

Ⓑ $\frac{1}{2}$

Ⓒ $\frac{3}{4}$

Ⓓ $\frac{2}{8}$

Part B: What is the probability of spinning it a second time and getting a 12 or greater both times?

19. A toy store owner sells action figures. He buys each one from the manufacture for $4.10. He has labor and other costs of $1.35 per action figure. He wants to make a 32% profit on each one. How much does he need to sell them for?

_____

20. Angle A and Angle B are complementary. Angle B measures 28°. What is the measure of Angle A?

Ⓐ 62°

Ⓑ 92°

Ⓒ 72°

Ⓓ 152°

21. Which of the following describes *all* requirements of similar polygons?

Ⓐ Similar polygons have congruent corresponding angles and proportional corresponding sides

Ⓑ Similar polygons have congruent corresponding angles and congruent corresponding sides

Ⓒ Similar polygons have proportional corresponding sides

Ⓓ Similar polygons have congruent corresponding angles

22. How far apart are the two points on the graph below? Each square represents 3 feet.

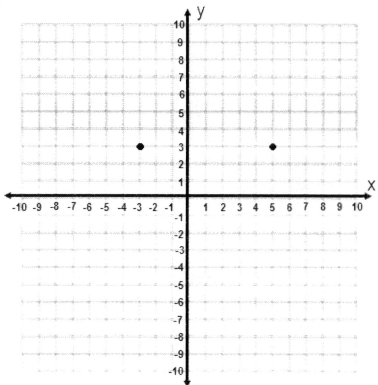

Ⓐ 15 feet

Ⓑ 22 feet

Ⓒ 21 feet

Ⓓ 24 feet

23. Eric is able to dribble the soccer ball down $\frac{2}{3}$ of the field in $\frac{2}{5}$ of a minute. How long will it take him to dribble the whole field?

Ⓐ 36 seconds

Ⓑ 24 seconds

Ⓒ 30 seconds

Ⓓ 32 seconds

24. Kaleb has a bike rim that is 18 inches in diameter. He puts a tire on it that is 2 inches thick. What is the circumference of the tire?

Ⓐ 20π

Ⓑ 22π

Ⓒ 24π

Ⓓ 18π

25. Given the trapezoid shown below, which of the following vertices represent the reflection of the trapezoid across the *y*-axis?

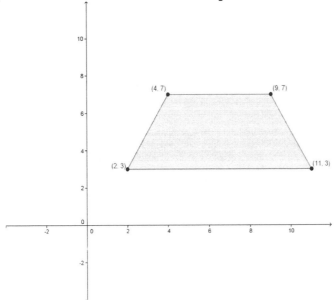

Ⓐ $(-4, -7), (-9, -7), (-2, -3), (-11, -3)$

Ⓑ $(-4, 7), (-9, 7), (-2, 3), (-11, 3)$

Ⓒ $(7, -4), (7, -9), (3, -2), (3, -11)$

Ⓓ $(4, -7), (9, -7), (2, -3), (11, -3)$

26. A regular heptagon has each side length equal to 9.2 cm. Which of the following is the best estimate for the perimeter of the heptagon?

Ⓐ 60 cm

Ⓑ 63 cm

Ⓒ 54 cm

Ⓓ 70 cm

27. A parallelogram has two bases, each equal to 18 cm, and a height of 8 cm. What is the area of the parallelogram?

Ⓐ 288 cm²

Ⓑ 72 cm²

Ⓒ 144 cm²

Ⓓ 96 cm²

28. Which of the following represents the area of the triangle shown below?

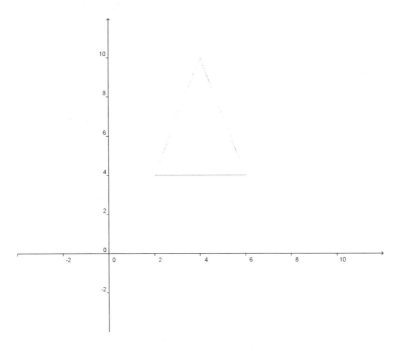

Ⓐ 8 square units

Ⓑ 9 square units

Ⓒ 10 square units

Ⓓ 12 square units

29. Judith purchased a box from the U.S. Postal Service with dimensions of 12 inches by 8 inches by 6 inches. How many cubic inches of space inside the box does she have available for use?

30. A pothole has a radius of 9 inches. Which of the following best represents the distance around the pothole?

Ⓐ 14.13 inches

Ⓑ 28.26 inches

Ⓒ 42.39 inches

Ⓓ 56.52 inches

31. What is the area of a trapezoid with base lengths of 7 cm and 10 cm and a height of 5 cm?

Ⓐ 85 cm²

Ⓑ 42.5 cm²

Ⓒ 28 cm²

Ⓓ 8.5 cm²

32. What is the sample space when rolling two standard dice?

Ⓐ 18

Ⓑ 6

Ⓒ 12

Ⓓ 36

33. What is the sample space when flipping a coin 9 times?

Ⓐ 256

Ⓑ 4,096

Ⓒ 512

Ⓓ 1,028

34. Kevin spins a spinner with 8 sections labeled 1 through 8. He also flips a coin. What is the probability he will land on a number less than 5 and get tails?

Ⓐ $\frac{7}{8}$

Ⓑ $\frac{1}{4}$

Ⓒ $\frac{5}{16}$

Ⓓ $\frac{1}{2}$

35. A box contains 8 yellow marbles, 9 orange marbles, and 1 green marble. What is the probability that Ann pulls out a yellow marble, replaces it, and then pulls a green marble?

    Ⓐ $\frac{4}{153}$

    Ⓑ $\frac{1}{2}$

    Ⓒ $\frac{4}{9}$

    Ⓓ $\frac{2}{81}$

36. The number of flights a flight attendant made per month is represented by the line graph below.

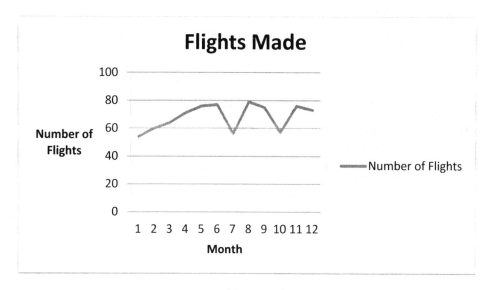

What is the range in the number of flights the flight attendant made?

    Ⓐ 20

    Ⓑ 25

    Ⓒ 29

    Ⓓ 32

37. Aubrey planted fruit trees on her farm. The number of each type of tree planted is shown in the table below.

Type of Tree	Number of Trees
Apple Tree	8
Peach Tree	18
Fig Tree	12
Pear Tree	3

Which circle graph represents the percentage of each type of tree planted?

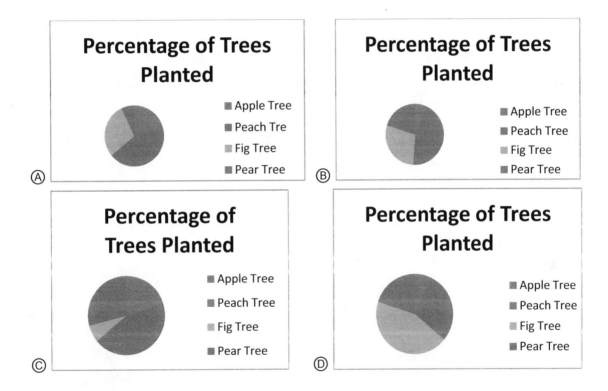

Ⓐ

Percentage of Trees Planted
- Apple Tree
- Peach Tre
- Fig Tree
- Pear Tree

Ⓑ

Percentage of Trees Planted
- Apple Tree
- Peach Tree
- Fig Tree
- Pear Tree

Ⓒ

Percentage of Trees Planted
- Apple Tree
- Peach Tree
- Fig Tree
- Pear Tree

Ⓓ

Percentage of Trees Planted
- Apple Tree
- Peach Tree
- Fig Tree
- Pear Tree

38. Chandler wishes to examine the median house value in his new hometown. Which graphical representation will most clearly indicate the median?

Ⓐ Box-and-whisker plot

Ⓑ Stem-and-leaf plot

Ⓒ Line plot

Ⓓ Bar graph

39. Part A: The number of long distance minutes Amanda used per week for business purposes is shown in the table below.

Week	Number of Minutes
1	289
2	255
3	322
4	291
5	306
6	302
7	411
8	418

What is the median number of long distance minutes she used?

_____

Part B: How much more is the mean than the median?

_____

40. A university reported the number of incoming freshmen from 2002 to 2011. The data is shown in the table below.

Year	Number of Incoming Freshmen
2002	7,046
2003	7,412
2004	6,938
2005	7,017
2006	7,692
2007	8,784
2008	7,929
2009	7,086
2010	8,017
2011	8,225

Based on the 10-year sample of data, which of the following represents the approximate average number of incoming freshmen?

Ⓐ 7,618

Ⓑ 7,615

Ⓒ 7,621

Ⓓ 7,624

Answers and Explanations

1. $\frac{14}{4}$: First you will want to convert all of the numbers to a decimal so they will be easier to compare. $\frac{14}{4} = 3.5, 3.41, \pi = 3.14, 3\frac{3}{8} = 3.375$. Once they are all in decimal form you can see that 3.5 or $\frac{14}{4}$ is the biggest.

2. B: To find miles per hour just divide the number of miles by the number of hours. In this case 3 hours and 20 minutes is equal to $3\frac{1}{3}$ hours. 1510 divided by $3\frac{1}{3}$ is approximately 453 mph.

3. B: The order of operations states that multiplication and division, as they appear from left to right in the expression, should be completed following the evaluation of exponents. Therefore, after evaluating the squared number, that value should be multiplied by 2.

4. C: The following proportion can be used to solve the problem: $\frac{35}{3} = \frac{x}{18}$, where x represents the number of mature trees. Solving for x gives $3x = 630$, which simplifies to $x = 210$.

5. D: The order of operations requires evaluation of the expression inside the parentheses as a first step. Thus, the expression can be re-written as $-8^2 + 8 \times 4 + 7$. Next, the integer with the exponent must be evaluated. Doing so gives $-64 + 8 \times 4 + 7$. The order of operations next requires all multiplications and divisions to be computed as they appear from left to right. Thus, the expression can be written as $-64 + 32 + 7$. Finally, the addition may be computed as it appears from left to right. The expression simplifies to $-32 + 7$, or -25.

6. A: Jason's number can be determined by writing the following expression: $\sqrt{2x - 4}$, where x represents Amy's number. Substitution of 20 for x gives $\sqrt{2(20) - 4}$, which simplifies to $\sqrt{36}$, or 6. Thus, Jason's number is 6. Jason's number can also be determined by working backwards. If Jason's number is the square root of 4 less than 2 times Amy's number, Amy's number should first be multiplied by 2 with 4 subtracted from that product and the square root taken of the resulting difference.

7. Part A: Since the line that divides the shaded and non shaded region runs from corner to corner it cuts the rectangle in half. This means you can just find the area of the rectangle and divide by 2. However there are two smaller rectangles like this. So, if you take half of each that is the same as one whole rectangle. The area of the rectangle is 5 cm times 12 cm which is 60 cm. Since you would divide by 2 to get the area of one but then multiply back by 2 to get the area of both there is no need to do either. The area of the shaded region is 60 sq cm.

Part B: B: As mentioned in Part A, since the line that divides them cuts the rectangle in half they are the same size.

8. 23: First find the relationship between x and y. When $x=0$ then $y=-2$, so this means that the equation is will have a -2 in it. If you add 2 back to all of the y numbers then you can see that they are the squares of the x's. So the relationship is $y = x^2 - 2$. Then you can just plug in to find the when $x=5, y=23$.

9.$\frac{5}{6}$: First plug 8 in for y to get $6x + 4 = 2(8) - 7$. Then solve for x. $6x + 4 = 16 - 7, 6x = 5, x = \frac{5}{6}$.

10. B: Based on the company's charge per half of an acre, the original charge is equal to 25×4, or $100, since there are 4 half-acres in 2 acres. With the discount of 20%, the following expression can be used to determine the final charge: $x - 0.20x$, where x represents the original charge. Substitution of 100 for x gives $100 - 0.20(100)$, which equals $100 - 20$, or 80. Thus, the company will charge $80.

11. B: The original price was $290,600 ($278,000 + $12,600). In order to determine the percentage of reduction, the following equation can be written: $\$12,600 = \$290,600x$, which simplifies to $x \approx 0.04$, or 4%. Thus, the percentage of reduction was approximately 4%.

12. $.47: The total was $5.17 and the 4 apples cost $3.76. So, $5.17-$3.76=$1.41 that was spent on bananas. Since there were 3 bananas, divide $1.41 by 3 to get $.47 per banana.

13. $51.20: First find what Melanie makes in a week. She makes $12 an hour times 40 hours, so she makes $480. Then take off 15% percent for her taxes. $480(.85)=$408. Next find what Lynn makes in one week. She makes $14 an hour times 40 hours, so she makes $560. Then take 18% off for her taxes. $560(.82)=$459.20. Then subtract what Melanie makes from what Lynn makes to find out how much more Lynn makes. $459.20-$408= $51.20.

14. B: The volume of a cone can be determined by using the formula $V = \frac{1}{3}\pi r^2 h$. Substitution of the radius and volume into the formula gives $150.72 = \frac{1}{3}\pi(4)^2 h$, which simplifies to $150.72 = \frac{1}{3}\pi 16h$. Division of each side of the equation by $\frac{1}{3}\pi 16$ gives $h = 9$. Thus, the height of the cone is 9 cm.

15. 3: The range of the numbers is the difference between the largest and smallest numbers in a set of numbers. In this case each tick mark on the number line represents $\frac{1}{2}$. The smallest number plotted is $5\frac{1}{2}$ and the largest number is $8\frac{1}{2}$. The range is 3.

16. $\frac{1}{2}, 1\frac{5}{8}$: $2\frac{1}{2} - \frac{3}{2}$ is already given as $1\frac{1}{8}$. Inside the parentheses convert the $\frac{2}{3}$ to $\frac{4}{6}$. Then you can do $\frac{4}{6} - \frac{1}{6} = \frac{3}{6} = \frac{1}{2}$. The first space is $\frac{1}{2}$. Then $1\frac{1}{8} + \frac{1}{2} = 1\frac{5}{8}$. The second space is $1\frac{5}{8}$.

17. B: Point T is equal to $2\frac{2}{3}$ and S is equal to 4. Perform all of the equations to figure out which one produces a number that is not on the number line. $2T-S = 2(2\frac{2}{3})-4 = 1\frac{1}{3}$ which is not on the number line.

18. Part A: C: There are a total of 8 spaces on the spinner all of equal size. There are 6 spaces that are 12 or greater. $\frac{6}{8} = \frac{3}{4}$.

Part B: $\frac{9}{16}$: The first time the probability was $\frac{3}{4}$ and the second time the probability is also $\frac{3}{4}$. The probability of it happening both times though is $\frac{3}{4} \times \frac{3}{4}$ which is $\frac{9}{16}$.

19. $7.02: If he buys each toy for $4.10 and then has another $1.30 in it, then he has a total of $5.40 in each toy. He marks it up 30% to sell it so his sales price is $5.40(1.30)=$7.02.

20. A: Complementary angles sum to 90 degrees. Since Angle B measures 28°, Angle A measures 90° − 28°, or 62°.

21. A: Similar polygons must have congruent corresponding angles and proportional corresponding sides. Both requirements must be fulfilled in order to declare similarity in polygons.

22. D: One point is at (-3, 3) and the other is at (5, 3). So, they are at a distance of 8. Since each square is equal to 3 feet they are 24 feet apart.

23. A: If it takes him $\frac{2}{5}$ of a minute to dribble $\frac{2}{3}$ of the field then divide by 2 to get $\frac{1}{5}$ of a minute for $\frac{1}{3}$ of the field. Then you can multiply by 3 to get $\frac{3}{3}$ of the field in $\frac{3}{5}$ of a minute. $\frac{3}{5}$ of a minute equals 36 seconds.

24. C: If he puts a tire on that is 2 inches thick then that adds 4 inches to the overall diameter. Now the radius is 12 inches. The formula for circumference is $2\pi r$. 2 times 12π is 24π.

25. B: A reflection of a figure across the y-axis is achieved by finding the additive inverse of each x-value. The y-values will not change. Therefore, the vertices of the reflected figure are (−4, 7), (−9, 7), (−2, 3), and (−11, 3).

26. B: A regular heptagon has equal side lengths. Thus, an estimate for the perimeter can be computed by rounding the given side length and multiplying by 7 (the number of sides of a heptagon); 9.2 can be rounded to 9, and 9 × 7 =63. Thus, the best estimate for the perimeter of the heptagon is 63 cm.

27. C: The area of a parallelogram can be calculated using the formula $A = bh$. The length of the base of the parallelogram is 18 cm, and the height is 8 cm. Thus, the area is equal to 18×8 cm^2, or 144 cm^2.

28. D: The given triangle has a base equal to 4 units and a height equal to 6 units. Thus, the area of the triangle is equal to $\frac{1}{2}(4)(6)$ square units, or 12 square units.

29. The correct answer is **576**. The box is a rectangular prism, and the amount of available space inside the box is synonymous with the volume of the box. The volume of a rectangular prism is calculated by finding the product of the length, width, and height. Thus, the volume of the box is equal to 12 in × 8 in × 6 in, or 576 cubic inches.

30. D: The distance around the pothole indicates the circumference of the pothole. The circumference of a circle can be determined by using the formula $C = \pi d$, where C represents the circumference and d represents the diameter. The diameter of the pothole is 18 inches (9 × 2). Substituting a diameter of 18 inches and 3.14 for the value of pi gives the following: $C = 3.14(18)$, or 56.52. Thus, the distance around the pothole is equal to 56.52 inches.

31. B: The area of a trapezoid can be found by using the formula $A = \frac{1}{2}(b_1 + b_2)h$, where b_1 and b_2 represent the lengths of the bases and h represents the height of the trapezoid. Substituting the given base lengths and height reveals the following: $A = \frac{1}{2}(7 + 10)5$, which equals 42.5. Thus, the area of the trapezoid is 42.5 cm^2.

32. D: The sample space of rolling each die is 6. Thus, the sample space of rolling two dice is equal to the product of the sample spaces. 6 × 6 = 36; therefore, the sample space is equal to 36.

33. C: Flipping a coin one time has a sample space equal to 2, i.e., T or H. Flipping a coin 2 times has a sample space equal to 4, i.e., TT, HH, TH, HT. Flipping a coin 3 times has a sample space of 8, i.e., TTT, HHH, THT, HTH, TTH, HHT, THH, HTT. Notice that 2 is equal to 2^1, 4 is equal to 2^2, and 8 is equal to 2^3. The sample space of flipping a coin 9 times is equal to 2^9, or 512.

34. B: The events are independent since the spin of a spinner does not have an effect on the outcome of the flip of a coin. The probability of two independent events can be found using the formula $P(A \text{ and } B) = P(A) \cdot P(B)$. The probability of landing on a number less than 5 is $\frac{4}{8}$ since there are 4 possible numbers less than 5 (1, 2, 3, and

4). The probability of getting tails is $\frac{1}{2}$. The probability of landing on a number less than 5 and getting tails is $\frac{4}{8} \cdot \frac{1}{2}$, which equals $\frac{4}{16}$, or $\frac{1}{4}$.

35. D: The events are independent since Ann replaces the first marble drawn. The probability of two independent events can be found using the formula $P(A \text{ and } B) = P(A) \cdot P(B)$. The probability of pulling out a yellow marble is $\frac{8}{18}$. The probability of pulling out a green marble after the yellow marble has been replaced is $\frac{1}{18}$. The probability that Ann pulls out a yellow marble and then a green marble is $\frac{8}{18} \cdot \frac{1}{18}$, which equals $\frac{8}{324}$, which reduces to $\frac{2}{81}$.

36. B: The line graph shows the largest number of flights made during a month as 79 with the smallest number of flights made during a month as 54. The range is equal to the difference between the largest number of flights and smallest number of flights, i.e., 79 – 54 = 25. Therefore, the range is equal to 25.

37. A: The percentages of each type of tree are as follows: Apple tree – 20%; Peach tree – 44%; Fig tree – 29%, and Pear tree – 7%. The circle graph for Choice A accurately represents these percentages.

38. A: The median can be determined using any of the given graphical representations. However, a box-and-whiskers plot actually includes a line drawn for the median, thus clearly indicating the value of the median.

39. The correct answer is **304.** The median number of minutes can be determined by listing the number of minutes in order from least to greatest and calculating the average of the two middle values. The number of minutes can be written in ascending order as 255, 289, 291, 302, 306, 322, 411, and 418. The two middle values are 302 and 306. The average of these values can be determined by writing $\frac{302+306}{2}$, which equals 304. Thus, the median number of minutes is 304.

40. B: The average number (or mean) of incoming freshmen can be calculated by summing the numbers of incoming freshmen and dividing by the total number of years (or 10). Thus, the mean can be calculated by evaluating $\frac{76,146}{10}$, which equals 7,614.6. Since a fraction of a person cannot occur, the mean can be rounded to 7,615 freshmen.

Success Strategies

The most important thing you can do is to ignore your fears and jump into the test immediately. Do not be overwhelmed by any strange-sounding terms. You have to jump into the test like jumping into a pool—all at once is the easiest way.

Make Predictions

As you read and understand the question, try to guess what the answer will be. Remember that several of the answer choices are wrong, and once you begin reading them, your mind will immediately become cluttered with answer choices designed to throw you off. Your mind is typically the most focused immediately after you have read the question and digested its contents. If you can, try to predict what the correct answer will be. You may be surprised at what you can predict.

Quickly scan the choices and see if your prediction is in the listed answer choices. If it is, then you can be quite confident that you have the right answer. It still won't hurt to check the other answer choices, but most of the time, you've got it!

Answer the Question

It may seem obvious to only pick answer choices that answer the question, but the test writers can create some excellent answer choices that are wrong. Don't pick an answer just because it sounds right, or you believe it to be true. It MUST answer the question. Once you've made your selection, always go back and check it against the question and make sure that you didn't misread the question and that the answer choice does answer the question posed.

Benchmark

After you read the first answer choice, decide if you think it sounds correct or not. If it doesn't, move on to the next answer choice. If it does, mentally mark that answer choice. This doesn't mean that you've definitely selected it as your answer choice, it just means that it's the best you've seen thus far. Go ahead and read the next choice. If the next choice is worse than the one you've already selected, keep going to the next answer choice. If the next choice is better than the choice you've already selected, mentally mark the new answer choice as your best guess.

The first answer choice that you select becomes your standard. Every other answer choice must be benchmarked against that standard. That choice is correct until proven otherwise by another answer choice beating it out. Once you've decided that no other answer choice seems as good, do one final check to ensure that your answer choice answers the question posed.

Valid Information

Don't discount any of the information provided in the question. Every piece of information may be necessary to determine the correct answer. None of the information in the question is there to throw you off (while the answer choices will certainly have information to throw you off). If two seemingly unrelated topics are discussed, don't ignore either. You can be confident there is a relationship, or it wouldn't be included in the question, and you are probably going to have to determine what is that relationship to find the answer.

Avoid "Fact Traps"

Don't get distracted by a choice that is factually true. Your search is for the answer that answers the question. Stay focused and don't fall for an answer that is true but irrelevant. Always go back to the question and make sure you're choosing an answer that actually answers the question and is not just a true statement. An answer can be factually correct, but it MUST answer the question asked. Additionally, two answers can both be seemingly correct, so be sure to read all of the answer choices, and make sure that you get the one that BEST answers the question.

Milk the Question

Some of the questions may throw you completely off. They might deal with a subject you have not been exposed to, or one that you haven't reviewed in years. While your lack of knowledge about the subject will be a hindrance, the question itself can give you many clues that will help you find the correct answer. Read the question carefully and look for clues. Watch particularly for adjectives and nouns describing difficult terms or words that you don't recognize. Regardless of whether you completely understand a word or not, replacing it with a synonym, either provided or one you more familiar with, may help you to understand what the questions are asking. Rather than wracking your mind about specific detailed information concerning a difficult term or word, try to use mental substitutes that are easier to understand.

The Trap of Familiarity

Don't just choose a word because you recognize it. On difficult questions, you may not recognize a number of words in the answer choices. The test writers don't put "make-believe" words on the test, so don't think that just because you only recognize all the words in one answer choice that that answer choice must be correct. If you only recognize words in one answer choice, then focus on that one. Is it correct? Try your best to determine if it is correct. If it is, that's great. If not, eliminate it. Each word and answer choice you eliminate increases your chances of getting the question correct, even if you then have to guess among the unfamiliar choices.

Eliminate Answers

Eliminate choices as soon as you realize they are wrong. But be careful! Make sure you consider all of the possible answer choices. Just because one appears right, doesn't mean that the next one won't be even better! The test writers will usually put more than one good answer choice for every question, so read all of them. Don't worry if you are stuck between two that seem right. By getting down to just two remaining possible choices, your odds are now 50/50. Rather than wasting too much time, play the odds. You are guessing, but guessing wisely because you've been able to knock out some of the answer choices that you know are wrong. If you are eliminating choices and realize that the last answer choice you are left with is also obviously wrong, don't panic. Start over and consider each choice again. There may easily be something that you missed the first time and will realize on the second pass.

Tough Questions

If you are stumped on a problem or it appears too hard or too difficult, don't waste time. Move on! Remember though, if you can quickly check for obviously incorrect answer choices, your chances of guessing correctly are greatly improved. Before you completely give up, at least try to knock out a couple of possible answers. Eliminate what you can and then guess at the remaining answer choices before moving on.

Brainstorm

If you get stuck on a difficult question, spend a few seconds quickly brainstorming. Run through the complete list of possible answer choices. Look at each choice and ask yourself, "Could this answer the question satisfactorily?" Go through each answer choice and consider it independently of the others. By systematically going through all possibilities, you may find something that you would otherwise overlook. Remember though that when you get stuck, it's important to try to keep moving.

Read Carefully

Understand the problem. Read the question and answer choices carefully. Don't miss the question because you misread the terms. You have plenty of time to read each question thoroughly and make sure you understand what is being asked. Yet a happy medium must be attained, so don't waste too much time. You must read carefully, but efficiently.

Face Value

When in doubt, use common sense. Always accept the situation in the problem at face value. Don't read too much into it. These problems will not require you to make huge leaps of logic. The test writers aren't trying to throw you off with a cheap trick. If you have to go beyond creativity and make a leap of logic in order to have an answer choice answer the question, then you should look at the other answer choices. Don't overcomplicate the problem by creating theoretical relationships or explanations that will warp time or space. These are normal problems rooted in reality. It's just that the applicable relationship or explanation may not be readily apparent and you have to figure things out. Use your common sense to interpret anything that isn't clear.

Prefixes

If you're having trouble with a word in the question or answer choices, try dissecting it. Take advantage of every clue that the word might include. Prefixes and suffixes can be a huge help. Usually they allow you to determine a basic meaning. Pre- means before, post- means after, pro - is positive, de- is negative.

From these prefixes and suffixes, you can get an idea of the general meaning of the word and try to put it into context. Beware though of any traps. Just because con- is the opposite of pro-, doesn't necessarily mean congress is the opposite of progress!

Hedge Phrases

Watch out for critical hedge phrases, led off with words such as "likely," "may," "can," "sometimes," "often," "almost," "mostly," "usually," "generally," "rarely," and "sometimes." Question writers insert these hedge phrases to cover every possibility. Often an answer choice will be wrong simply because it leaves no room for exception. Unless the situation calls for them, avoid answer choices that have definitive words like "exactly," and "always."

Switchback Words

Stay alert for "switchbacks." These are the words and phrases frequently used to alert you to shifts in thought. The most common switchback word is "but." Others include "although," "however," "nevertheless," "on the other hand," "even though," "while," "in spite of," "despite," and "regardless of."

New Information

Correct answer choices will rarely have completely new information included. Answer choices typically are straightforward reflections of the material asked about and will directly relate to the question. If a new piece of information is included in an answer choice that doesn't even seem to relate to the topic being asked about, then that answer choice is likely incorrect. All of the information needed to answer the question is usually provided for you in the question. You should not have to make guesses that are unsupported or choose answer choices that require unknown information that cannot be reasoned from what is given.

Time Management

On technical questions, don't get lost on the technical terms. Don't spend too much time on any one question. If you don't know what a term means, then odds are you aren't going to get much further since you don't have a dictionary. You should be able to immediately recognize whether or not you know a term. If you don't, work with the other clues that you have—the other answer choices and terms provided—but don't waste too much time trying to figure out a difficult term that you don't know.

Contextual Clues

Look for contextual clues. An answer can be right but not the correct answer. The contextual clues will help you find the answer that is most right and is correct. Understand the context in which a phrase or statement is made. This will help you make important distinctions.

Don't Panic

Panicking will not answer any questions for you; therefore, it isn't helpful. When you first see the question, if your mind goes blank, take a deep breath. Force yourself to mechanically go through the steps of solving the problem using the strategies you've learned.

Pace Yourself

Don't get clock fever. It's easy to be overwhelmed when you're looking at a page full of questions, your mind is full of random thoughts and feeling confused, and the clock is ticking down faster than you would like. Calm down and maintain the pace that you have set for yourself. As long as you are on track by monitoring your pace, you are guaranteed to have enough time for yourself. When you get to the last few minutes of the test, it may seem like you won't have enough time left, but if you only have as many questions as you should have left at that point, then you're right on track!

Answer Selection

The best way to pick an answer choice is to eliminate all of those that are wrong, until only one is left and confirm that is the correct answer. Sometimes though, an answer choice may immediately look right. Be careful! Take a second to make sure that the other choices are not equally obvious. Don't make a hasty mistake. There are only two times that you should stop before checking other answers. First is when you are positive that the answer choice you have selected is correct. Second is when time is almost out and you have to make a quick guess!

Check Your Work

Since you will probably not know every term listed and the answer to every question, it is important that you get credit for the ones that you do know. Don't miss any questions through careless mistakes. If at all possible, try to take a second to look back over your answer selection and make sure you've selected the correct answer choice and haven't made a costly careless mistake (such as marking an answer choice that you didn't mean to mark). The time it takes for this quick double check should more than pay for itself in caught mistakes.

Beware of Directly Quoted Answers

Sometimes an answer choice will repeat word for word a portion of the question or reference section. However, beware of such exact duplication. It may be a trap! More than likely, the correct choice will paraphrase or summarize a point, rather than being exactly the same wording.

Slang

Scientific sounding answers are better than slang ones. An answer choice that begins "To compare the outcomes..." is much more likely to be correct than one that begins "Because some people insisted..."

Extreme Statements

Avoid wild answers that throw out highly controversial ideas that are proclaimed as established fact. An answer choice that states the "process should used in certain situations, if..." is much more likely to be correct than one that states the "process should be discontinued completely." The first is a calm rational statement and doesn't even make a definitive, uncompromising stance, using a hedge word "if" to provide wiggle room, whereas the second choice is a radical idea and far more extreme.

Answer Choice Families

When you have two or more answer choices that are direct opposites or parallels, one of them is usually the correct answer. For instance, if one answer choice states "x increases" and another answer choice states "x decreases" or "y increases," then those two or three answer choices are very similar in construction and fall into the same family of answer choices. A family of answer choices consists of two or three answer choices, very similar in construction, but often with directly opposite meanings. Usually the correct answer choice will be in that family of answer choices. The "odd man out" or answer choice that doesn't seem to fit the parallel construction of the other answer choices is more likely to be incorrect.

How to Overcome Test Anxiety

The very nature of tests caters to some level of anxiety, nervousness, or tension, just as we feel for any important event that occurs in our lives. A little bit of anxiety or nervousness can be a good thing. It helps us with motivation, and makes achievement just that much sweeter. However, too much anxiety can be a problem, especially if it hinders our ability to function and perform.

"Test anxiety," is the term that refers to the emotional reactions that some test-takers experience when faced with a test or exam. Having a fear of testing and exams is based upon a rational fear, since the test-taker's performance can shape the course of an academic career. Nevertheless, experiencing excessive fear of examinations will only interfere with the test-taker's ability to perform and chance to be successful.

There are a large variety of causes that can contribute to the development and sensation of test anxiety. These include, but are not limited to, lack of preparation and worrying about issues surrounding the test.

Lack of Preparation

Lack of preparation can be identified by the following behaviors or situations:
- Not scheduling enough time to study, and therefore cramming the night before the test or exam
- Managing time poorly, to create the sensation that there is not enough time to do everything
- Failing to organize the text information in advance, so that the study material consists of the entire text and not simply the pertinent information
- Poor overall studying habits

Worrying, on the other hand, can be related to both the test taker, or many other factors around him/her that will be affected by the results of the test. These include worrying about:
- Previous performances on similar exams, or exams in general
- How friends and other students are achieving
- The negative consequences that will result from a poor grade or failure

There are three primary elements to test anxiety. Physical components, which involve the same typical bodily reactions as those to acute anxiety (to be discussed below). Emotional factors have to do with fear or panic. Mental or cognitive issues concerning attention spans and memory abilities.

Physical Signals

There are many different symptoms of test anxiety, and these are not limited to mental and emotional strain. Frequently there are a range of physical signals that will let a test taker know that he/she is suffering from test anxiety. These bodily changes can include the following:

- Perspiring
- Sweaty palms
- Wet, trembling hands
- Nausea
- Dry mouth
- A knot in the stomach
- Headache
- Faintness
- Muscle tension
- Aching shoulders, back and neck
- Rapid heart beat
- Feeling too hot/cold

To recognize the sensation of test anxiety, a test-taker should monitor him/herself for the following sensations:

- The physical distress symptoms as listed above
- Emotional sensitivity, expressing emotional feelings such as the need to cry or laugh too much, or a sensation of anger or helplessness
- A decreased ability to think, causing the test-taker to blank out or have racing thoughts that are hard to organize or control.

Though most students will feel some level of anxiety when faced with a test or exam, the majority can cope with that anxiety and maintain it at a manageable level. However, those who cannot are faced with a very real and very serious condition, which can and should be controlled for the immeasurable benefit of this sufferer.

Naturally, these sensations lead to negative results for the testing experience. The most common effects of test anxiety have to do with nervousness and mental blocking.

Nervousness

Nervousness can appear in several different levels:

- The test-taker's difficulty, or even inability to read and understand the questions on the test
- The difficulty or inability to organize thoughts to a coherent form
- The difficulty or inability to recall key words and concepts relating to the testing questions (especially essays)

- The receipt of poor grades on a test, though the test material was well known by the test taker

Conversely, a person may also experience mental blocking, which involves:
- Blanking out on test questions
- Only remembering the correct answers to the questions when the test has already finished.

Fortunately for test anxiety sufferers, beating these feelings, to a large degree, has to do with proper preparation. When a test taker has a feeling of preparedness, then anxiety will be dramatically lessened.

The first step to resolving anxiety issues is to distinguish which of the two types of anxiety are being suffered. If the anxiety is a direct result of a lack of preparation, this should be considered a normal reaction, and the anxiety level (as opposed to the test results) shouldn't be anything to worry about. However, if, when adequately prepared, the test-taker still panics, blanks out, or seems to overreact, this is not a fully rational reaction. While this can be considered normal too, there are many ways to combat and overcome these effects.

Remember that anxiety cannot be entirely eliminated, however, there are ways to minimize it, to make the anxiety easier to manage. Preparation is one of the best ways to minimize test anxiety. Therefore the following techniques are wise in order to best fight off any anxiety that may want to build.

To begin with, try to avoid cramming before a test, whenever it is possible. By trying to memorize an entire term's worth of information in one day, you'll be shocking your system, and not giving yourself a very good chance to absorb the information. This is an easy path to anxiety, so for those who suffer from test anxiety, cramming should not even be considered an option.

Instead of cramming, work throughout the semester to combine all of the material which is presented throughout the semester, and work on it gradually as the course goes by, making sure to master the main concepts first, leaving minor details for a week or so before the test.

To study for the upcoming exam, be sure to pose questions that may be on the examination, to gauge the ability to answer them by integrating the ideas from your texts, notes and lectures, as well as any supplementary readings.

If it is truly impossible to cover all of the information that was covered in that particular term, concentrate on the most important portions, that can be covered very well. Learn these concepts as best as possible, so that when the test comes, a goal can be made to use these concepts as presentations of your knowledge.

In addition to study habits, changes in attitude are critical to beating a struggle with test anxiety. In fact, an improvement of the perspective over the entire test-taking experience can actually help a test taker to enjoy studying and therefore improve the overall experience. Be certain not to overemphasize the significance of the grade - know that the result of the test is neither a reflection of self worth, nor is it a measure of intelligence; one grade will not predict a person's future success.

To improve an overall testing outlook, the following steps should be tried:
- Keeping in mind that the most reasonable expectation for taking a test is to expect to try to demonstrate as much of what you know as you possibly can.
- Reminding ourselves that a test is only one test; this is not the only one, and there will be others.
- The thought of thinking of oneself in an irrational, all-or-nothing term should be avoided at all costs.
- A reward should be designated for after the test, so there's something to look forward to. Whether it be going to a movie, going out to eat, or simply visiting friends, schedule it in advance, and do it no matter what result is expected on the exam.

Test-takers should also keep in mind that the basics are some of the most important things, even beyond anti-anxiety techniques and studying. Never neglect the basic social, emotional and biological needs, in order to try to absorb information. In order to best achieve, these three factors must be held as just as important as the studying itself.

Study Steps

Remember the following important steps for studying:
- Maintain healthy nutrition and exercise habits. Continue both your recreational activities and social pass times. These both contribute to your physical and emotional well being.
- Be certain to get a good amount of sleep, especially the night before the test, because when you're overtired you are not able to perform to the best of your best ability.
- Keep the studying pace to a moderate level by taking breaks when they are needed, and varying the work whenever possible, to keep the mind fresh instead of getting bored.
- When enough studying has been done that all the material that can be learned has been learned, and the test taker is prepared for the test, stop studying and do something relaxing such as listening to music, watching a movie, or taking a warm bubble bath.

There are also many other techniques to minimize the uneasiness or apprehension that is experienced along with test anxiety before, during, or even after the examination. In fact, there are a great deal of things that can be done to stop anxiety from interfering with lifestyle and performance. Again, remember that anxiety will not be eliminated entirely, and it shouldn't be. Otherwise that "up" feeling for exams would not exist, and most of us depend on that sensation to perform better than usual. However, this anxiety has to be at a level that is manageable.

Of course, as we have just discussed, being prepared for the exam is half the battle right away. Attending all classes, finding out what knowledge will be expected on the exam, and knowing the exam schedules are easy steps to lowering anxiety. Keeping up with work will remove the need to cram, and efficient study habits will eliminate wasted time. Studying should be done in an ideal location for concentration, so that it is simple to become interested in the material and give it complete attention. A method such as SQ3R (Survey, Question, Read, Recite, Review) is a wonderful key to follow to make sure that the study habits are as effective as possible, especially in the case of learning from a textbook. Flashcards are great techniques for memorization. Learning to take good notes will mean that notes will be full of useful information, so that less sifting will need to be done to seek out what is pertinent for studying. Reviewing notes after class and then again on occasion will keep the information fresh in the mind. From notes that have been taken summary sheets and outlines can be made for simpler reviewing.

A study group can also be a very motivational and helpful place to study, as there will be a sharing of ideas, all of the minds can work together, to make sure that everyone understands, and the studying will be made more interesting because it will be a social occasion.

Basically, though, as long as the test-taker remains organized and self confident, with efficient study habits, less time will need to be spent studying, and higher grades will be achieved.

To become self confident, there are many useful steps. The first of these is "self talk." It has been shown through extensive research, that self-talk for students who suffer from test anxiety, should be well monitored, in order to make sure that it contributes to self confidence as opposed to sinking the student. Frequently the self talk of test-anxious students is negative or self-defeating, thinking that everyone else is smarter and faster, that they always mess up, and that if they don't do well, they'll fail the entire course. It is important to decreasing anxiety that awareness is made of self talk. Try writing any negative self thoughts and then disputing them with a positive statement instead. Begin self-encouragement as though it was a friend speaking. Repeat positive statements to help reprogram the mind to believing in successes instead of failures.

Helpful Techniques

Other extremely helpful techniques include:
- Self-visualization of doing well and reaching goals
- While aiming for an "A" level of understanding, don't try to "overprotect" by setting your expectations lower. This will only convince the mind to stop studying in order to meet the lower expectations.
- Don't make comparisons with the results or habits of other students. These are individual factors, and different things work for different people, causing different results.
- Strive to become an expert in learning what works well, and what can be done in order to improve. Consider collecting this data in a journal.
- Create rewards for after studying instead of doing things before studying that will only turn into avoidance behaviors.
- Make a practice of relaxing - by using methods such as progressive relaxation, self-hypnosis, guided imagery, etc - in order to make relaxation an automatic sensation.
- Work on creating a state of relaxed concentration so that concentrating will take on the focus of the mind, so that none will be wasted on worrying.
- Take good care of the physical self by eating well and getting enough sleep.
- Plan in time for exercise and stick to this plan.

Beyond these techniques, there are other methods to be used before, during and after the test that will help the test-taker perform well in addition to overcoming anxiety.

Before the exam comes the academic preparation. This involves establishing a study schedule and beginning at least one week before the actual date of the test. By doing this, the anxiety of not having enough time to study for the test will be automatically eliminated. Moreover, this will make the studying a much more effective experience, ensuring that the learning will be an easier process. This relieves much undue pressure on the test-taker.

Summary sheets, note cards, and flash cards with the main concepts and examples of these main concepts should be prepared in advance of the actual studying time. A topic should never be eliminated from this process. By omitting a topic because it isn't expected to be on the test is only setting up the test-taker for anxiety should it actually appear on the exam. Utilize the course syllabus for laying out the topics that should be studied. Carefully go over the notes that were made in class, paying special attention to any of the issues that the professor took special care to emphasize while lecturing in class. In the textbooks, use the chapter review, or if possible, the chapter tests, to begin your review.

It may even be possible to ask the instructor what information will be covered on the exam, or what the format of the exam will be (for example, multiple choice, essay, free form, true-false). Additionally, see if it is possible to find out how many questions will be on the test. If a review sheet or sample test has been offered by the professor, make good use of it, above anything else, for the preparation for the test. Another great resource for getting to know the examination is reviewing tests from previous semesters. Use these tests to review, and aim to achieve a 100% score on each of the possible topics. With a few exceptions, the goal that you set for yourself is the highest one that you will reach.

Take all of the questions that were assigned as homework, and rework them to any other possible course material. The more problems reworked, the more skill and confidence will form as a result. When forming the solution to a problem, write out each of the steps. Don't simply do head work. By doing as many steps on paper as possible, much clarification and therefore confidence will be formed. Do this with as many homework problems as possible, before checking the answers. By checking the answer after each problem, a reinforcement will exist, that will not be on the exam. Study situations should be as exam-like as possible, to prime the test-taker's system for the experience. By waiting to check the answers at the end, a psychological advantage will be formed, to decrease the stress factor.

Another fantastic reason for not cramming is the avoidance of confusion in concepts, especially when it comes to mathematics. 8-10 hours of study will become one hundred percent more effective if it is spread out over a week or at least several days, instead of doing it all in one sitting. Recognize that the human brain requires time in order to assimilate new material, so frequent breaks and a span of study time over several days will be much more beneficial.

Additionally, don't study right up until the point of the exam. Studying should stop a minimum of one hour before the exam begins. This allows the brain to rest and put things in their proper order. This will also provide the time to become as relaxed as possible when going into the examination room. The test-taker will also have time to eat well and eat sensibly. Know that the brain needs food as much as the rest of the body. With enough food and enough sleep, as well as a relaxed attitude, the body and the mind are primed for success.

Avoid any anxious classmates who are talking about the exam. These students only spread anxiety, and are not worth sharing the anxious sentimentalities.

Before the test also involves creating a positive attitude, so mental preparation should also be a point of concentration. There are many keys to creating a positive attitude. Should fears become rushing in, make a visualization of taking the exam, doing well, and seeing an A written on the paper. Write out a list of affirmations that will bring a feeling of confidence, such as "I am doing well in my English class," "I studied well and know my material," "I enjoy this class." Even if the affirmations aren't believed at first, it sends a positive message to the subconscious which will result in an alteration of the overall belief system, which is the system that creates reality.

If a sensation of panic begins, work with the fear and imagine the very worst! Work through the entire scenario of not passing the test, failing the entire course, and dropping out of school, followed by not getting a job, and pushing a shopping cart through the dark alley where you'll live. This will place things into perspective! Then, practice deep breathing and create a visualization of the opposite situation - achieving an "A" on the exam, passing the entire course, receiving the degree at a graduation ceremony.

On the day of the test, there are many things to be done to ensure the best results, as well as the most calm outlook. The following stages are suggested in order to maximize test-taking potential:

- Begin the examination day with a moderate breakfast, and avoid any coffee or beverages with caffeine if the test taker is prone to jitters. Even people who are used to managing caffeine can feel jittery or light-headed when it is taken on a test day.
- Attempt to do something that is relaxing before the examination begins. As last minute cramming clouds the mastering of overall concepts, it is better to use this time to create a calming outlook.
- Be certain to arrive at the test location well in advance, in order to provide time to select a location that is away from doors, windows and other distractions, as well as giving enough time to relax before the test begins.
- Keep away from anxiety generating classmates who will upset the sensation of stability and relaxation that is being attempted before the exam.
- Should the waiting period before the exam begins cause anxiety, create a self-distraction by reading a light magazine or something else that is relaxing and simple.

During the exam itself, read the entire exam from beginning to end, and find out how much time should be allotted to each individual problem. Once writing the exam, should more time be taken for a problem, it should be abandoned, in order to begin another problem. If there is time at the end, the unfinished problem can always be returned to and completed.

Read the instructions very carefully - twice - so that unpleasant surprises won't follow during or after the exam has ended.

When writing the exam, pretend that the situation is actually simply the completion of homework within a library, or at home. This will assist in forming a relaxed atmosphere, and will allow the brain extra focus for the complex thinking function.

Begin the exam with all of the questions with which the most confidence is felt. This will build the confidence level regarding the entire exam and will begin a quality momentum. This will also create encouragement for trying the problems where uncertainty resides.

Going with the "gut instinct" is always the way to go when solving a problem. Second guessing should be avoided at all costs. Have confidence in the ability to do well.

For essay questions, create an outline in advance that will keep the mind organized and make certain that all of the points are remembered. For multiple choice, read every answer, even if the correct one has been spotted - a better one may exist.

Continue at a pace that is reasonable and not rushed, in order to be able to work carefully. Provide enough time to go over the answers at the end, to check for small errors that can be corrected.

Should a feeling of panic begin, breathe deeply, and think of the feeling of the body releasing sand through its pores. Visualize a calm, peaceful place, and include all of the sights, sounds and sensations of this image. Continue the deep breathing, and take a few minutes to continue this with closed eyes. When all is well again, return to the test.

If a "blanking" occurs for a certain question, skip it and move on to the next question. There will be time to return to the other question later. Get everything done that can be done, first, to guarantee all the grades that can be compiled, and to build all of the confidence possible. Then return to the weaker questions to build the marks from there.

Remember, one's own reality can be created, so as long as the belief is there, success will follow. And remember: anxiety can happen later, right now, there's an exam to be written!

After the examination is complete, whether there is a feeling for a good grade or a bad grade, don't dwell on the exam, and be certain to follow through on the reward that was promised...and enjoy it! Don't dwell on any mistakes that have been made, as there is nothing that can be done at this point anyway.

Additionally, don't begin to study for the next test right away. Do something relaxing for a while, and let the mind relax and prepare itself to begin absorbing information again.

From the results of the exam - both the grade and the entire experience, be certain to learn from what has gone on. Perfect studying habits and work some more on confidence in order to make the next examination experience even better than the last one.

Learn to avoid places where openings occurred for laziness, procrastination and day dreaming.

Use the time between this exam and the next one to better learn to relax, even learning to relax on cue, so that any anxiety can be controlled during the next exam. Learn how to relax the body. Slouch in your chair if that helps. Tighten and then relax all of the different muscle groups, one group at a time, beginning with the feet and then working all the way up to the neck and face. This will ultimately relax the muscles more than they were to begin with. Learn how to breathe deeply and comfortably, and focus on this breathing going in and out as a relaxing thought. With every exhale, repeat the word "relax."

As common as test anxiety is, it is very possible to overcome it. Make yourself one of the test-takers who overcome this frustrating hindrance.

Additional Bonus Material

Due to our efforts to try to keep this book to a manageable length, we've created a link that will give you access to all of your additional bonus material.

Please visit http://www.mometrix.com/bonus948/fsag7math to access the information.